Photo by] *Henry Irving.*
CORSICAN PINE

PITMAN'S COMMON COMMODITIES
AND INDUSTRIES

GUMS & RESINS

THEIR OCCURRENCE, PROPERTIES
AND USES

BY

ERNEST J. PARRY,
B.Sc., F.I.C., F.C.S.

LONDON: SIR ISAAC PITMAN & SONS, LTD.
BATH, MELBOURNE AND NEW YORK

COMMON COMMODITIES AND INDUSTRIES SERIES

Each book in crown 8vo, cloth, with many illustrations, charts, etc., 2/6 net

TEA. By A. Ibbetson
COFFEE. By B. B. Keable
SUGAR. By Geo. Martineau, C.B.
OILS. By C. Ainsworth Mitchell, B.A., F.I.C.
WHEAT. By Andrew Millar
RUBBER. By C. Beadle and H. P. Stevens, M.A., Ph.D., F.I.C.
IRON AND STEEL. By C. Hood
COPPER. By H. K. Picard
COAL. By Francis H. Wilson, M.Inst., M.E.
TIMBER. By W. Bullock
COTTON. By R. J. Peake
SILK. By Luther Hooper
WOOL. By J. A. Hunter
LINEN. By Alfred S. Moore
TOBACCO. By A. E. Tanner
LEATHER. By K. J. Adcock
KNITTED FABRICS. By J. Chamberlain and J. H. Quilter
CLAYS. By Alfred B. Searle
PAPER. By Harry A. Maddox
SOAP. By William A. Simmons, B.Sc. (Lond.), F.C.S.
THE MOTOR INDUSTRY. By Horace Wyatt, B.A.
GLASS AND GLASS MAKING. By Percival Marson
GUMS AND RESINS. By E. J. Parry, B.Sc., F.I.C., F.C.S.
THE BOOT AND SHOE INDUSTRY. By J. S. Harding
GAS AND GAS MAKING. By W. H. Y. Webber
FURNITURE. By H. E. Binstead
COAL TAR. By A. R. Warnes
PETROLEUM. By A. Lidgett
SALT. By A. F. Calvert
ZINC. By T. E. Lones, M.A., LL.D., B.Sc.
PHOTOGRAPHY. By Wm. Gamble
ASBESTOS. By A. Leonard Summers
SILVER. By Benjamin White
CARPETS. By Reginald S Brinton
PAINTS AND VARNISHES. By A. S. Jennings
CORDAGE AND CORDAGE HEMP AND FIBRES. By T. Woodhouse and P. Kilgour
ACIDS AND ALKALIS. By G. H. J. Adlam
ELECTRICITY. By R. E. Neale, B.Sc., Hons.
ALUMINIUM. By Captain G. Mortimer
GOLD. By Benjamin White.
BUTTER AND CHEESE. By C. W. Walker-Tisdale and Jean Jones.
THE BRITISH CORN TRADE. By A. Barker.
LEAD. By Dr. J. A. Smythe.
STONES AND QUARRIES. By J. Allan Howe.
ENGRAVING. By T. W. Lascelles.

OTHERS IN PREPARATION

PREFACE

THE present little work is designed only to bring within the same covers the principal bodies known as resins, gums, and gum resins in a convenient form for reference.

Chemical information is, of course, absolutely necessary in dealing with this group of natural substances, but this has been kept down to its simplest possible limits, in order that this little handbook should be in keeping with the remainder of the series of which it forms a part.

It is obvious that, since thousands of trees yield resinous substances, a selective treatment is necessary, so that only the more important substances, or those which, for some reason or other, possess special interest, have been dealt with.

The terms gum, resin, and balsam are used in a more or less loose form popularly. The varnish resins are frequently spoken of as varnish gums, and so on. A simple, if incomplete, distinction, which will be easily understood, is that the true gums are more or less soluble in water, and insoluble in organic solvents, whilst the resins are insoluble in water, and more or less soluble in organic solvents.

CONTENTS

CHAP		PAGE
	PREFACE	iii
I.	THE TRUE OR PROPER RESINS	1
II.	THE GUM RESINS	30
III.	BALSAMS, MEDICINAL RESINS, ETC.	38
IV.	THE TRUE GUMS	78

ILLUSTRATIONS

		PAGE
CORSICAN PINE.		*Frontispiece*
AGATHIS ROBUSTA, THE QUEENSLAND KAURI TREE		3
THE SCOTS PINE		13
AN INDIAN PINE PLANTATION		31
MYROXYLON PEREIRAE, THE PERU BALSAM TREE		45
INDIAN PINE FOREST ON ROCKS		51
LIQUIDAMBAR STYRACIFLUA		55

GUMS AND RESINS

CHAPTER I

RESINS PROPER

Copal Resin

Copal resin, or gum copal, as it is frequently termed in commerce, comprises a number of different types of resin, some of recent, others of fossil origin, and to some extent found in nearly all tropical and subtropical countries throughout the world. The principal sources of the copal of commerce are East Africa, West Africa, the Dutch East Indies, certain islands in Polynesia, New Zealand, New Caledonia and the north-eastern portions of South America. Generally speaking, five principal types of copal are recognised in commerce, which are as follows—

1. East African copal, including Zanzibar, etc.
2. West African copal.
3. Manila copal.
4. Kauri copal, from New Zealand and New Caledonia.
5. South American Copal.

The East African product is collected in British, Portuguese, and (the late) German East Africa, and is usually sent thence to Zanzibar where it is sorted, cleaned and packed for export. It is known as Zanzibar animi or copal, and varies greatly in price, in normal times dust fetching about £30 per ton and fine grades over £300. The value of the exports from Zanzibar reaches about £15,000 per annum. The East African copals are fossil resins, being found principally in localities from which copal trees have disappeared. They are probably the product of species of *Trachylobium*.

The West African copals are obtained along the coastal regions of West Africa from Sierra Leone to the Portuguese Congo. The finer varieties are fossil or semi-fossil, and the poorer grades are derived from living trees. The best varieties are obtained from the Congo, Angola and Benguela; the medium qualities from Sierra Leone and Accra, and the low grades from the Niger districts. The trees which yield or have yielded these types are probably *Copaifera guibourtiana Cyanothryrsus ogea*, and *Daniella oblonga*. They are of far less value than East African copal, the best qualities, in normal times, being worth about £75 per ton.

Manila copal was at one time produced entirely in the Philippine Islands, but the same type of resin is now gathered in the Dutch East Indies and comes into commerce under the names Macassar, Pontianac, or Singapore copals. These copals are used for the cheaper, but still good, qualities of copal varnish.

Kauri copal, as the resin from New Zealand and New Caledonia is termed, is a fossil resin derived from the so-called Kauri pine, *Dammara Australis*. The finest varieties fetch about as much as the best East African copal. New Zealand exports about £500,000 to £600,000 worth of this resin per annum.

South American copal, which is the product of *Hymenoea* species, is derived principally from the living trees, but a certain amount of fossil resin is also collected. Commercially it is known as Demerara animi, and the best grades are worth about half as much as the best East African varieties.

AGATHIS ROBUSTA
(The Queensland Kauri Tree)

(*Baker & Smith, " The Pines of Australasia "*)

Copal resin varies much in appearance. It is a hard, brittle resin, vitreous and semi-transparent, and yellow to red in colour, often having a peculiar faceted or indented surface, technically known as "goose skin," which is especially characteristic of Zanzibar copal.

The principal virtue of copal resin is its hardness, on account of which copal varnish dries with a good hard surface, able to withstand considerable wear. It is, as is usually the case with the so-called hard "varnish gums," necessary to heat the resin until destructive decomposition sets in, and about 10 to 25 per cent. of its weight is lost in the form of water, gas and oil, before it becomes soluble in linseed oil and turpentine in order to convert it into varnish. The only "adulterants" of copal resin met with in commerce are the cheaper and softer varieties, which are sometimes mixed with the harder and more expensive grades. Common resin has been said to be an adulterant, but this is apocryphal.

Generally speaking, it may be taken for granted that, after allowing for colour, the value of a copal resin depends practically entirely on its hardness. The principal exception to this rule is Kauri copal, which, although it may be of low melting point, is so easily manipulated by the varnish manufacturer, that it is valued for certain purposes irrespective of its hardness.

The following table represents the general scale of hardness of the principal types of copal known—

       1. Zanzibar copal.
       2. Mozambique copal.
       3. Lindi copal.
       4. Red Angola copal.
       5. Pebble copal.

6. Sierra Leone copal (fossil).
7. Yellow Benguela copal.
8. White Benguela copal.
9. Cameroon copal.
10. Congo copal.
11. Manila copal.
12. White Angola copal.
13. Kauri copal.
14. Sierra Leone copal (living trees).
15. South American copals.

The actual constituents of copal resin are practically unknown. Numerous compounds have been reported as present, but their very high molecular weights, and the absence of characteristic derivatives, render the published details very dubious, and they must be accepted with reserve. For example, Tschirch and Stephan claim to have isolated from Zanzibar copal, about 80 per cent. of trachylolic acid, of the formula $C_{54} H_{85} O_3 (OH) (COOH)_2$, and other similar compounds. The author can find little evidence beyond the percentage results of an organic combustion, to support these formulae. It is, however, certain that all copals consist, like shellac, of a mixture of which the principal constituents are free acids and esters, and the determination of the acid and ester values gives useful information as to the purity and type of the resin examined. The following figures, which must not be considered exhaustive, have been found to cover the types of copal resin mentioned—

	Acid value.	Ester value.
Manila	120 to 130	45 to 55
Singapore	120 to 135	50 to 65
Kauri	65 to 85	30 to 40
Angola	50 to 90	50 to 80

The iodine value has also been determined on a

number of samples and varies as follows according to Worstall—

Kauri	74 to 170
Manila	104 to 148
Pontianac	119 to 142
Zanzibar	115 to 123
Mozambique	136
Madagascar	126
West African	122 to 143
Sierra Leone	102 to 105
Brazilian	123 to 134

DAMMAR RESIN

The various types of dammar resin are derived from a number of trees, of which the principal species are *Hopea, Shorea* and *Balanocarpus*. The greater part of the supply of dammar resin (" gum dammar ") is produced in the Federated Malay States, Sumatra, and other islands of the Dutch East Indies. The fact that dammar resin is fairly soluble in alcohol and turpentine, causes it to be used to a fairly considerable extent in the manufacture of the so-called " spirit varnishes " which are used for the varnishing of fabrics such as paper and certain cloths, and for indoor decorative work.

Dammar resin is not a fossil resin, all types being obtained from living trees.

The East Indian dammar resins, principally obtained from Java, Singapore and Padang are the principal varieties of any importance in the European markets. This East Indian dammar is in the form of stalactitic granules and lumps, of an almost white to yellowish colour, transparent, usually coated with dust, and having a conchoidal fracture. It is not so hard as copal, but is still what may be termed one of the hard varnish resins. In addition to the above described, there are certain dark resins, often nearly black, obtained from certain districts in India. They are, however, not of great commercial importance.

The finest dammar is the Batavian variety, which is valued on account of its fine colour, and because it forms a very clear solution with turpentine.

Nine samples of dammar resin, obtained from the Federated Malay States, were chemically examined in the Imperial Institute, and found to have the following characters—

	Melting point.	Ash.	Acid number.	Ester number.
(1)	90°C.	0·26%	45·3	1·4
(2)	94°	0·08%	72·0	0
(3)	87°	0·05%	38·5	0
(4)	180°	0·52%	33·0	1·3
(5)	185°	0·03%	72·0	0
(6)	92°	0·06%	33·0	0
(7)	200°	0·04%	46·5	0·2
(8)	97°	0·25%	38·5	0
(9)	190°	0·09%	55·0	0

Dammar resin is very soluble in alcohol, benzene, turpentine and carbon disulphide.

Colophony can be detected as an adulterant of dammar resin by digesting a small portion of the sample for half an hour with ammonia solution, and acidifying the filtered liquid with acetic acid. If colophony be present, a precipitate of resin acids is formed, whilst if the dammar be pure only a slight opalescence occurs, since only a minute amount of the acids of dammar resin is extracted by ammonia.

A sample of dammar resin known as "rock dammar" from Burma has been examined, and has been found to be of excellent quality, and fit to be classed as amongst the better varieties. It is very abundant in the South Tenasserim Division of Burma, and is derived from *Hopea odorata*. According to expert reports, it is suitable for crystal varnishes, and if it could be obtained in the fossilised state it is probable that the demand for it would be very large indeed.

This Burmese Rock dammar has the following characters—

Ash	=	0·55 to 0·68%
Saponification No.	=	31 to 37·1
Acid No.	=	31 to 31·5
Ester No.	=	0 to 5·6
Melting point	=	90° to 115°C.
Specific gravity	=	0·980 to 1·013

Gottlieb has recently described two types of recent dammar resin from mid-Borneo (*Arch. Pharm.* 1911, 249-701). Of these one is known as dammar daging and is probably identical with "Rose dammar." It is obtained from *Retinodendron rassak*. It forms bright yellowish-white pieces, some of them having a reddish tinge. It has the following characters—

Initial melting point	=	130°C.
Complete melting point	=	150°C.
Acid value	=	140–151
Saponification value	=	159–165
Soluble in alcohol	=	82%
,, petroleum ether	=	30%
,, carbon bisulphide	=	45%
,, benzene	=	25%

It consists almost entirely of resin acids with a small amount of neutral resenes.

The second resin is a recent fossil resin, of unknown origin. It consists largely of resin acids, but contains about 8 per cent. of an essential oil and a certain amount of bassorin.

ELEMI RESINS

The name "elemi" is in the ordinary way practically restricted to the somewhat soft aromatic oleo resinous body collected in the Philippine Islands from one or more species of *Canarium*, principally *Canarium luzonicum*. There are small quantities of other resins offered under the name elemi from time to time, but these are usually qualified by the name of the place from which they have been imported.

In the fresh condition elemi resin is a mixture of resin with a certain amount of essential oil, of pale colour, and either soft and somewhat viscid, when it is known as soft elemi, or hard and semi-crystalline, when less essential oil is present. The smell of elemi reminds one of a mixture of lemon and turpentine. According to Dieterich the following resins are properly described as true elemi resins—

1. Manila elemi, from *Canarium luzonicum*.
2. *Yucatan elemi, from Amyris plumieri.*
3. *Mexican elemi*, from *Amyris elemifera.*
4. Rio elemi, from various plants.
5. Brazilian elemi, from *Protium heptaphyllum.*
6. African elemi, from *Boswellia freriana.*
7. East Indian elemi, from *Canarium zephyrenum.*

Until quite recently the botanical origin of Manila elemi, which is the principal elemi of commerce, was a matter of speculation, but researches by the Bureau of Science of the United States of America have established the fact that the oleoresin is collected in the Philippines from *Canarium luzonicum*. The fresh oleoresin contains from 20 to 30 per cent. of an essential oil, which is composed mainly of hydrocarbons, of which the terpene phellandrene is the principal.

The non-volatile resin consists largely of two easily crystallisable compounds, known as *alpha*-amyrin and *beta*-amyrin.

At one time elemi resin was used to a considerable extent in medicine, as an ingredient in various ointments and plasters, but to-day this has entirely ceased, and it is principally employed in the preparation of printing inks, and, to a smaller extent, in certain types of varnish.

The characters of elemi resin differ according to the district in which it is produced. The following is a fair average of the figures obtained in the analysis of typical Manila elemi—

Soft Manila Elemi.

Volatile oil	=	15 to 20%
Ash	=	0·02 to 0·2%
Acid value	=	17 to 25
Ester value	=	7 to 25

Hard Manila Elemi.

Volatile oil	=	8 to 9%
Ash	=	0·2 to 1%
Acid value	=	15 to 28
Ester value	=	25 to 35

There is a resin very similar in character to ordinary elemi resin, which is produced in Dominica, probably from *Bursera gummifera*, the so-called gommier tree, and is known as gommier resin. It is of common occurrence in the forests of Dominica, and is collected and used by the natives for the manufacture of torches and incense. The resin exudes either from natural fissures or from cuts in the bark. It is at first an opaque whitish, viscous liquid, which soon dries into yellowish lumps of brittle resin. What little has reached the London market has been sold as " dry " or West Indian elemi.

The hard variety of this resin is completely soluble in alcohol and partially so in oil of turpentine, whilst the soft variety is completely soluble in turpentine oil, but only partially so in alcohol.

Elemi resins from Southern Nigeria and from Uganda have also been imported and examined, and if carefully collected and exported in a clean condition, it is probable that these elemis would find a ready market as a substitute for the Manila product.

Essential oil of elemi is fragrant, with an odour recalling those of fennel, lemon and turpentine, a sp. gr. of 0·87–0·91, and is dextrorotatory. The terpenes present consist chiefly of phellandrene and dipentene. The resin of elemi consists of two substances, one, brein, soluble in cold, and the other, amyrin, in hot alcohol, the former occurring in the larger proportion (60 p.c.). The one soluble in hot alcohol is left when the elemi is treated with cold alcohol, and can be obtained in the form of white crystals, to the extent of 20 per cent., by crystallisation from boiling alcohol. This resin is neutral. The water in which the elemi is distilled, retains two crystallisable substances soluble in water to which the name of bryoidin and breidine have been given. Manila elemi also contains a small quantity of a crystalline acid named elemic acid, the crystals of which are larger than those of the other crystalline bodies above mentioned.

Colophony or Common Resin

This resin is the cheapest of all commercial resins, and is more largely employed than any other. At present the principal source of supply is the United States, but considerable quantities are collected in France and Russia, and recent researches and developments indicate that there are vast possibilities for the industry in India. (See pp. 31 and 51.)

Colophony is the non-volatile portion of the oleoresinous exudation of various species of pine trees, which are to be found in enormous quantities in the producing regions. This oleoresinous exudation is known as crude turpentine, which, on steam distillation yields the oil of turpentine of commerce, leaving the rosin behind. Common rosin, or colophony, is used for numerous purposes in the arts, including the manufacture of very low grade varnishes, cheap household soaps, for the distillation of rosin spirit and rosin oil, and for the manufacture of metallic resinates, which are added to varnishes to assist rapid drying. Varnish made from colophony is of very low grade and weathers very badly: indeed, powdered rosin can usually be scratched with the finger from the varnished article. A small, but very important technical use for colophony has quite recently arisen in the manufacture of ester gums, as they are called. The colophony, being almost entirely of an acid nature, combines with the alcohol glycerine, and with certain other bodies containing hydroxy groups, forming a stable ester, or salt of the acid body present. These "ester gums" have been found to be far more useful than ordinary colophony for varnish manufacture, as the dried varnish weathers well, and cannot be scratched or removed with anything like the ease that colophony varnish can. To indicate the value of the rosin industry, we may draw attention to the imports of the year 1907 into the United Kingdom, when rosin to the value of £896,301 was brought into the country, of which no less than £693,065 came from the United States, and £136,092 from France.

The methods by which rosin is obtained from the pine tree vary to a certain extent, but the following description will fairly indicate the general principles underlying its production. In the United States the principal tree used for turpentine-tapping is *Pinus Australis*, but numerous other pines are also employed to a less extent. In the autumn and winter the trees are "boxed," that is, excavations of characteristic shape are made in the trunks of the trees, about 8 in. above the ground. These excavations are known as boxes and are so made as to

Photo by] [Henry Irving.

SCOTS PINE

hold from 5 to 10 lb. of the exudation. After allowing a few days after the boxing, the bark is cut away for about 3 ft. above the box and the wood is cut with grooves leading to the box so that the oleoresin shall collect there and not run away. The exudation of the crude turpentine commences about the following March and goes on till the end of August when it becomes very slow, and then finishes about the middle of October. The crude oleoresin is then baled out into barrels and conveyed to the stills and heated to drive off water. It is then distilled, and the volatile essential oil sold as oil of turpentine, and the non-volatile colophony or rosin is left in the still. As a rule, the " tapping " life of these trees is from five to eight years, after which they yield but little exudation.

In France the turpentine and rosin industry is practically confined to the Landes district and the principal tree used is *Pinus pinaster*. The crude oleoresin, known in France as the " gemme," exudes from the trees during the warm season, from March to October, from an incision made by the collector with an axe. This incision is known as the " carre," and is kept open by the removal of a thin slice once a week, and is gradually extended to a height of about 12 ft. from the ground. The tree is worked for one year and then left alone for two or three years, when a fresh incision is made, and the tree, by this means will yield oleoresin in payable quantity for a period of about forty years. It is then " bled to death," that is, worked by means of several incisions simultaneously, and so quite exhausted, before it is handed over to the tree fellers.

The Indian pine tree, which is known locally as " chir," is the *Pinus longifolia*, and the method of collecting the oleoresin is based on that in vogue in France. An initial cut, about 6 in. by 4 in. and 1 in.

deep is cut near the base of the tree, and slightly extended every week throughout the summer, until it is about 18 in. long by the end of the year. The oleoresin collects in a cup fixed at the base of this cut or " blaze," as it is called, and the contents are emptied periodically. Two classes of tapping are in use, (1) light tapping, and (2) heavy tapping. The latter system is carried out in the case of all trees due to be felled within five years, and consists in making as many blazes as possible, so that the tree is, as in the case of the French trees, bled to death, before being handed over to the feller.

Turpentine is also made to a large extent in Russia, but it is a different product to the above and the rosin industry is of much less importance.

At the present moment American and French rosins are the two commercial varieties, hardly any other ever being seen on the London market. French rosin is usually known as galipot, and American grades are lettered, for example A is nearly black, and W.W. is almost colourless (" water white").

Colophony consists almost entirely of a free acid, or mixture of free acids, known as abietic acid, possibly in the form of an anhydride, which is known chemically as a lactone. A small quantity of esters also exists, but considerably smaller than that found in most other resins. A good quality colophony is of a pale yellow colour, soft, easy to fracture with the fingers and practically transparent. On warming a distinct terebinethinate odour is noticeable. It is easily soluble in alcohol or in acetic acid, and in volatile and fixed oils. It is slightly heavier than water, its specific gravity being from 1·0450 to 1·085. It softens at about 75°C., and is completely melted at 120° to 135°. Being the cheapest resin of commerce, colophony is never adulterated. The analytical figures of typical colophony are as follows—

Specific gravity	=	1·0450 to 1·085
Acid value	=	150 to 175
Ester value	=	7 to 20
Iodine value	=	118 to 128
Unsaponifiable matter	=	4 to 9%

A useful quantitative test for colophony is the reaction known as the Storch-Morawski reaction. If a fragment of colophony be dissolved in acetic anhydride and the mixture allowed to cool, and the liquid filtered, the latter yields a fine reddish-violet coloration when sulphuric acid of specific gravity 1·53 is allowed to flow gently down the tube containing the acetic solution. The colour appears at the junction layer of the two liquids.

The so-called " driers " of the paint trade are prepared by melting colophony with the oxide of the metal, usually lead or manganese, or by the addition of a solution of a suitable metallic salt to an aqueous solution of the colophony in the form of its sodium salt. These resinate driers always contain a large excess of resin, as otherwise their action would be far too powerful for general use. To be satisfactory, the resinate driers must be completely soluble in linseed oil, and any insoluble metallic oxide is quite useless.

On dry distillation colophony yields the commercial products known as rosin spirit and rosin oil. The process is carried out in vertical cast-iron stills. On distillation gas and aqueous liquid are first driven off, and then follows a light, oily liquid, which boils between 80° and 250°C., and, when purified is known as rosin spirit. At about 290° to 310° rosin oil commences to distil over. The residue in the still consists of valuable pitch or of cokey matter, according to the length to which the distillation has gone. Rosin spirit is a pale or colourless oil of specific gravity about 0·850 to 0·880, and has been used to some extent as a substitute for turpentine.

Rosin oil is a viscid liquid varying in colour from a very pale yellow to dark brown. It is usually strongly fluorescent, but the "bloom," as it is called, can be removed by suitable treatment with dinitronaphthalene. Rosin oil has a specific gravity varying from 0·980 to 1·100, and consists principally of hydrocarbons with a small quantity of resin acids. It has a large use in the lubrication of machinery and wagon wheels, and when mixed with lime and petroleum oils form the axle-grease of commerce. It is also used, with or without linseed oil in the manufacture of printer's ink.

AMBER

Amber is the fossil resin derived from *Pinites succinifer* and is collected principally near the Prussian Baltic coast. Amber is the hardest known resin, being a brittle substance breaking with a conchoidal fracture. The colour varies from pale yellow to dark brown and even almost black. Some varieties are nearly transparent, others are cloudy and even opaque, the markings on the polished amber often being very beautiful. It polishes well, and possesses the character of being very easily electrified negatively when melted.

Amber is used to a considerable extent for the manufacture of ornamental articles: for example, cigar holders, the mouthpieces of pipes, etc., etc. Formerly the darker varieties and smaller pieces were used to make amber varnish, which, for fine work, such as carriage varnishing, is about the finest and hardest that can be made.

Very little of the varnish labelled " amber varnish "

to-day, however, contains any amber. In the working of amber, a considerable quantity of shavings is produced. These are amalgamated again with the assistance of enormous pressure. Sometimes other resins, such as copal, are worked in. This product, of course, has nothing like the wearing qualities of the natural amber, and is known as "imitation amber."

The varieties of amber recognised in commerce are as follows—

1. Succinite. This is the most important variety and forms pale yellow or yellowish-brown brittle lumps, either transparent or opaque and melting at 250° to 300°C.

2. Gedanite is so called "soft amber." This is of a whitish-yellow colour, easily fractured, and melting at 150° to 180°C.

3. Glessite is of a darker colour and is usually opaque. It melts at 250° to 300°.

4. Beckerite. This is a brown, opaque variety.

When amber is used for varnish manufacture, it is first melted and a certain amount of volatile oil is distilled off. This is the true oil of amber or *oleum succini*. True oil of amber has a specific gravity of about 0·950 and an optical rotation of $+15°$ to $+25°$. Nearly all the oil of amber of commerce, however, is not, in fact, distilled from amber at all, but is the product of the distillation of other resins, including colophony.

Pure amber has the following characters—

Acid value = 15 to 35
Ester value = 70 to 95

Genuine amber can easily be distinguished from the imitation amber described above by means of polarised light. When examined between crossed nicol prisms, it shows only very faint colours, whereas the lack of homogeneity in the manufactured article causes it to show brilliant colours.

THE ACAROID RESINS

There are two distinct varieties of acaroid resin, or gum acaroid as it is sometimes called, the red and the yellow. They are derived from species of *Xanthorrhoea*, especially *Xanthorrhoea Australis* and *Xanthorrhoea hastilis*.

The red resin is commonly known in Australia, whence both varieties are obtained, as " grass-tree gum," and occurs as small red brown dusty lumps with a lustrous fracture. The yellow variety is often termed Botany Bay gum, and comes principally from Tasmania.

They are cheap resins, not much more expensive than common colophony, as a substitute for which they are, to some extent, employed. They enter into the composition of a certain amount of cheap sealing wax, spirit varnishes for coating metals, and, in alkaline solution, for sizeing paper. The red variety is also used in spirit solution for staining wood a mahogany colour.

According to J. C. Umney (*Perfumery and Essential Oil Record*, 1915, 212) the light variety of the resin obtained from New South Wales is obtained from *Xanthorrhoea hastilis*, and the dark variety from *Xanthorrhoea arborea*. The former is the more aromatic, and is of particular interest because it contains an appreciable amount of benzoic acid. It possesses a certain amount of perfume value, and can, up to a point, be used as a substitute for benzoin, storax or tolu. On burning it emits a fragrant odour resembling that of gum benzoin, and might therefore be used as an incense resin. The percentage of benzoic acid present varies from 4·6 to 7·2 per cent.

This resin also possesses particular interest at the present time on account of the fact that, when heated with nitric acid, it yields a considerable quantity of picric acid. If the resin be destructively distilled in an iron retort, a large amount of oil is obtained, the heavy oil resembling phenol, and the light oil having the general character of benzene. It is obvious, therefore, that its general constitution must be fairly closely related to the closed chain hydrocarbons, or these products would not result.

An interesting account of the collection and properties of the resins of various species of *Xanthorrhoea* was published in the *Pharmaceutical Journal*, vol. xxi, p. 906 from which the following is abstracted—

" The stems of the grass trees are chopped down broken up into convenient pieces and allowed to fall into a sheet. A stout stick or flail completes the work of disintegration. The substance is then passed through a sieve, the ligneous portions being thus removed, and a gentle breeze is sufficient to winnow what has passed through the sieve. After a bush-fire has passed over grass-trees the heat causes the resin to run into more or less spherical masses, which become darker in colour. The heat of the sun is sometimes sufficient to produce this effect. As a rule, the commercial product is in small pieces, almost in powder, or in a friable mass of particles. When boiled with water a small quantity of tannin and colouring matter is extracted, and crystals of benzoic acid separate on cooling."

There is also a red acaroid resin obtained in the West Indies—principally in the Bahamas. It is, however paler in colour than the Australian variety.

The following method has been described by Rabs for identifying acaroid resin in the presence of copal shellac and colophony. A small amount of the material to be examined is heated with ten to twenty drops of nitric acid until nitrous fumes are copiously evolved. When cold, the residue is dissolved in alcohol and ten to twenty drops of a 5 per cent. solution of ferric chloride are added to the intensely red liquid. A brown to brownish-black coloration accompanied by a cloudy appearance in the liquid denotes the presence of acaroid resin.

The bulk of acaroid resin appears to consist of an ester of para-coumaric acid, with a certain amount of resin alcohols.

L. E. Andés (*Chem. Rev. Fett. Harz-Ind.*, 1909, 16, 160) gives the following interesting account of the acaroid resins—

" Of the various species of *Xanthorrhoea*, *X. drummondii* (W. Australia) is reputed to afford most resin, a single tree yielding an average of 23 kilos. of a yellow resin. *X. tateana* (S. Australia and Kangaroo Island) furnishes a ligneous, vesiculated, readily friable and odourless resin. The mass is dark red; the powder is yellowish and imparts a blood-red colour to hot water. Petroleum ether extracts 1 per cent. of a colourless odourless resin; strong alcohol dissolves it entirely, forming a fiery red solution, which deposits crystals of benzoic acid on evaporation. *X. hastilis* (N.S.W. and Queensland) produces a resin of sweetish odour, resembling that of benzoin; it is readily friable, the powder resembling gamboge and undergoing change of colour when exposed to light. It melts in boiling water, rendering the latter turbid and yellow. Petroleum ether extracts 1 per cent. of a pleasant-smelling substance; alcohol dissolves 94 per cent., and the solution affords feathery crystals of benzoic acid on evaporation. The purified resin melts at 97·7°C. Another sample, showing a lower melting point, yielded to petroleum ether 2 per cent. of a faintly coloured viscous body, probably composed of essential oils and resin. *X. arborea* (N.S.W. and Queensland) furnishes compact pieces mixed with leaves; the colour of the product varies from purple brown to carmine red. It forms a readily friable powder, of the colour of raw sienna, and tastes like benzoin. Petroleum ether extracts 8 per cent.; alcohol, 92 per cent.; the alcohol extract deposits crystals of benzoic acid, but in less quantity than the other *Xanthorrhoea* resins. *X. Australis* (Tasmania and Victoria) affords irregular-shaped spheroidal masses of friable resin of a dark red colour, in the fused state resembling dragon's blood. Its alcohol solution is clearer than those of the resins of other species of *Xanthorrhoea*."

SANDARAC RESIN

This resin is derived from the North West African tree *Callitris quadrivalis*. A number of Australian *Callitris* species, which have recently been examined, yield the so-called "pine-gum" or Australian sandarac. The African sandarac, chiefly exported from Mogador is in the form of yellowish lumps, dusty on the outside and easily pulverised. It is used principally for the preparation of spirit varnishes.

The resin contains a considerable amount of free acids, notably pimaric acid, and yields a small quantity of an essential oil containing dextropinene and a diterpene. It melts at about 160°C. It is soluble in alcohol, ether, acetone and numerous essential oils. Its acid value varies from 90 to 154, generally from 140 to 154.

The Sandarac resins of the Australian *Callitris* species deserve special attention, as being products of our own Empire which have recently been investigated by

Messrs. Baker & Smith, of the New South Wales Technological Museum. With some species of the tree the resin is found in larger tears and masses than is common with the African resin, a peculiarity noticeable particularly in the resin of *Callitris culcavata*. That of *Callitris arenosa* very closely resembles the African variety. The chemical characters of the Australian resins appear to agree with those of the African variety.[1]

Mastic Resin

Mastic resin is the product of *Pisticia lentiscus*, one of the Anacardiacae, found abundantly on the shores of the Mediterranean and especially cultivated in Chios, where the finest quality is obtained. A certain amount of mastic is also obtained from India. The ordinary mastic resin of commerce is found in granular fragments about a quarter to three-quarters of an inch long. The resin is hard with a conchoidal fracture, and is more or less waxy when chewed owing to its low melting point and the absence of brittleness when warm. The granules are somewhat opaque, and of a yellowish or yellowish-green colour. Mastic is never adulterated in the lump form, but occasionally sandarac or colophony has been added to powdered mastic resin. It is soluble in alcohol and in most organic solvents, with the exception of petroleum ether. Mastic is used to a certain extent in lithographic work and also for pale-coloured spirit varnishes.

Mastic resin softens at under 100°C. and melts at under 110°. The following analytical values have been recorded by Williams—

	I.	II.
Acid value	50	56
Ester value	23	23·1

[1] From Baker & Smith's Report on the Pines of Australasia.

	I	*II.*
Saponification value	73	79
Ash	0·21%	0·14%
Moisture	0·97%	1·40%

Colophony and sandarac can be detected, when used as adulterants, by their high acid values.

GUAIACUM RESIN

This resin is obtained from the wood of *Guaiacum officinale* and, probably, *Guaiacum sanctum*, natives of Tropical America. It occurs in the form of masses of a greenish-black colour, and also in small fragments known as tears. The guaiacum in lumps is often "false packed" and contains very large quantities of woody fibre and similar foreign material.

The only use of guaiacum resin, so far as the author is aware, is in medicine, as it finds considerable employment in the treatment of gouty and rheumatic affections.

It is adulterated when in powder form with powdered resins of less value, especially powdered colophony. On account of its use as a medicine its analytical examination is a matter of considerable importance. Pure guaiacum resin is soluble in alcohol to the extent of 87 to 96 per cent., and in ether to the extent of 55 to 75 per cent. The ash of the best varieties rarely exceeds 3·5 per cent., or, in indifferent qualities 7 per cent. The acid value of genuine guaiacum resin, together with the ester and saponification values, are given by Dieterich as follows—

Acid value = 46–53
Ester value = 121–139
Saponification value = 167–192

In the detection of colophony, the Storch-Morawski reaction described under colophony may be used. According to Hirschsohn colophony, or the so-called Peruvian guaiacum resin may be detected by adding bromine solution to a chloroformic solution of the sample. If pure a blue colour results: with adulterated guaiacum, the coloration is red.

Dragon's Blood

The resin known as dragon's blood or "Sanguis draconis" is the product of a large species of rattan palm, *Daemonorops draco*, a native of the islands of the Indian Archipelago, but principally produced in Sumatra and Borneo. This variety is known as "Palm dragon's blood." Socotra dragon's blood is the produce of *Dracoena cinnabari*, and is produced in South Eastern Asia, Socotra and the West Indies.

The best dragon's blood is imported in the form of cylindrical rolls about 1 in thick and 10 to 12 in long, wrapped in palm leaves Sometimes small balls are imported, but these are not now frequently met with. The so-called "Socotra dragon's blood" is imported in conchoidal lumps.

The resin is of a deep red colour, as its name implies, and is used principally as a stain. The resin as imported varies enormously in quality, and a given parcel is usually sorted into several grades by an expert sorter, and the so-graded parcels ground into powder of varying qualities. The resin is quite dark, red in colour, and opaque. When powdered it yields a crimson-coloured powder soluble in alcohol.

The constituents of dragon's blood are obscure resin acids, alcohols, and esters, of very complex constitution.

Dragon's blood is used solely as a stain, or red spirit varnish. It is entirely soluble in alcohol, except for the presence of a small amount of mechanical impurities, such as vegetable fibre or sand.

According to Dieterich, palm dragon's blood, which is the finest variety, alone answers to the "draco-alban" test, which thus discriminates between Sumatra and Socotra dragon's blood. The test is as follows: ten grammés of powdered dragon's blood are extracted with ether, and the ethereal extract poured into absolute alcohol. If the resin be Sumatra dragon's blood, a white resinous precipitate is formed, which is not the case with Socotra resin.

The saponification value of dragon's blood varies from 135 to 145. Colophony can be detected by the Storch-Morawski reaction.

SHELLAC

Few resins are of greater commercial importance than shellac. It is used for the manufacture of varnish and polish to a very large extent, forming the basis of the French polish of the furniture maker, and of the best sealing-wax. It is also used to impart stiffness to numerous soft materials, such as hats, *crêpe* and similar articles, the fabric being soaked in a solution of the resin, and the solvent driven off by heat, leaving the substance stiffened with the deposited shellac.

The shellac industry is one of the more important of the minor branches of commerce in India. The resin is secreted by the so-called Lac insect, *Tachardia lacca*, one of the family of the Coccidae, or wax insects, which subsists upon the vegetable sap of certain trees which they suck up by piercing a hole in the tissues of their host by means of a sharp proboscis. The principal tree upon which the insects live is *Butea frondosa*, known locally as the palas tree. The insects swarm twice or even thrice a year, when the twigs of the trees infested will frequently assume a reddish colour due to the countless masses of insect larvae moving all over them. Those that survive and become fixed to the trees at once commence to transform, by a digestive process, the plant sap which they suck up, and to exude from their bodies the resinous secretion with which they ultimately become entirely incrusted.

The twigs, with their incrustations of lac are brought down from the forest into the factories, where they are broken into short lengths by hand or machinery, and in this condition, the crude product is known as " stick lac." This is next crushed by a mill, and the fragments of wood and dirt sifted out as far as possible, leaving the fragments of " seedlac," which are then washed and are ready to be converted into the " leaf lac " of commerce which is what is usually understood by the single word " shellac," or into the cakes known as " garnet lac " and " button lac."

In certain cases, where a particular colour is desired, a little yellow sulphide of arsenic is added, and in the great majority of cases, a little common American rosin. It was at one time believed that about 3 per cent. of rosin assisted the manufacture considerably, but to-day, rosin must be regarded merely as an adulterant. It has become customary in the shellac trade to allow an admixture of up to 3 per cent. of this adulterant, and the standard grade, which is known as ⟨TN⟩ may be accepted as almost invariably containing .2 to 3 per cent. of rosin. Garnet lac, a deep coloured form of shellac, manufactured from the dark coloured seed lac in thick cakes, is recognised in the trade as containing 10 per cent. of adulterating rosin, whilst button lac is sold in various standards, from absolutely pure, down to types containing 40 or even 50 per cent. of rosin.

The various forms of shellac are manufactured by melting the washed seed lac, mixed with the various quantities of rosin or not, as the case may be, and squeezing the melted substance through filter bags. Considerable skill is required to know just when the lac is sufficiently " cooked." A red dye accompanies this resinous secretion of the insect, which used to form an important article of commerce, but lac dye is now only a waste product, as its value has been killed by the coal-tar dyes.

The commerce in shellac is very considerable, and there is a large amount of speculation in the article. Within recent years, the price of <TN> shellac has been as low as about 55s. per cwt., and, at the top of a speculative boom, as high as 400s. In the year 1900, the exports of shellac, exclusive of button lac and seed lac, reached the high figure of 195,239 cwt. Calcutta is the chief centre of the world's trade in shellac, and India practically enjoys a monopoly of the supply. Burma and Madras send large supplies to Calcutta, but Assam also sends a fairly good supply. Whilst it is true that nearly every province in India possesses a lac supply, the product is mostly consumed locally, except in the case of the Central Provinces, Bengal, Assam and Burma.

Shellac consists of about 4 per cent. of wax, closely related in its composition to beeswax, with a small amount of colouring matter, the residue being a resin soluble in alcohol, or in solutions of alkali or borax. This last-named property enables a varnish to be inexpensively prepared, by dissolving the shellac in a solution of borax. Those boot and shoe dressings which are applied with a sponge to the leather are usually such solutions, mixed with a small amount of coal-tar black dye. The resin consists of a considerable proportion of free acid, which has been named aleuritic acid, together with a large amount of esters, which are compounds of bodies of an alcoholic nature, with acids.

The analysis of shellac is of considerable importance, as the natives are frequently unable to resist adding more rosin than is considered legitimate. The subject was in a quite chaotic condition until 1903, when the methods now in use were worked out and published by the author. (E. J. Parry, *Chemist and Druggist*, 31st Jan., 1903.) A most useful quantitative test for detecting the presence of added rosin is as follows: dissolve the sample in a little alcohol, and pour the solution into excess of water. The shellac is then precipitated in a very finely divided condition. This is collected and dried on filter paper. It is then triturated in a mortar with a little petroleum ether for five or ten minutes, and the petroleum ether filtered off. The rosin, if present, is dissolved by the petroleum ether, and if this be shaken with a 1 per cent. watery solution of copper acetate, it will assume an emerald green colour, due to the formation of a compound of the rosin with the copper, which is soluble in petroleum ether. For the quantitative determination of rosin in shellac, the principle suggested by the author in 1903 is now universally adopted. It was found that shellac formed a combination with iodine, to the extent of, on the average, 9 per cent. of its own weight, whilst rosin combines in a similar manner with 125 per cent. of its own weight. By a careful determination of the amount of iodine with which a given sample will combine, it is easy to calculate, from the above figures, the percentages of the two ingredients present.

CHAPTER II

GUM RESINS

THE gum resins are exudations from plants, which consist of a mixture of gums, soluble in water, with a small amount of essential oils, and a considerable proportion of true resinous matter.

Only a few of these are of commercial importance. They are used, generally speaking, in medicine, and usually occur in the form of lumps, or when of the finest quality, in small globules known as tears. Pure gum resins, when treated with water yield a certain amount of gum, with a little sugar or closely allied carbohydrates, and other inert water-soluble extractives.

GAMBOGE

Gum gamboge is found on the market in the form of cylindrical pipes from 1 to 2 in. in diameter, and of bright yellow to orange-yellow colour. It is sometimes found in cakes with a conchoidal fracture. It is the product of *Garcinia morella*, and is produced almost entirely in Siam, the Ceylon variety being comparatively rare. The gum resin is produced in the form of a juice, which is prepared by breaking off the leaves and shoots of the tree, when the juice issues in drops from the broken surfaces. These are collected, and the moisture evaporated, and when the juice has attained the required consistence, it is rolled into cylinders. Sometimes it is poured into the hollow parts of bamboos. The cylindrical rolls of gamboge contract on drying, and so usually become hollow in the centre. The name gamboge—or

INDIAN PINE PLANTATION
From "Indian Forest Memoirs" (R. S. Troup)

camboge—is derived from the province of Cambodia, whence it was originally procured.

The principal employment of gamboge is as a pigment, and to a certain extent in medicine, as it acts as an energetic cathartic.

Gamboge is frequently adulterated, either with farinaceous matter, dextrine, or colophony. According to Williams, pure gamboge has the following characters—

Acid value	80·6
Ester value	67·2
Saponification value	147·8
Ash	0·48%
Moisture	3·7%

ASAFOETIDA

This gum resin is the exudation of various species of *Ferula*, principally *Ferula scorodorma*, *Ferula narthex* and *Ferula foetida*. It is principally collected in and near Persia, the bulk of it being brought to Bombay, whence it is exported to Europe.

Asafoetida is a foul-smelling substance, commonly known as "Devils' Dung," and is used to a small extent in medicine for human beings, and, especially in America, as a veterinary medicine, as a remedy for certain diseases in horses. It is also used to a small extent as a flavouring, a minute amount being present in a good many of the sauces sold as condiments in this country.

Asafoetida occurs in the form of dry lumps of all sizes, from "tears" upwards, and also as sticky masses of all colours and consistencies. It is frequently mixed with other resinous material, but as the exact botanical origin of the commercial product is still a matter of some doubt, it is difficult to declare a given sample adulterated, unless so coarse an adulterant as woody fibre, starch or sand be present.

According to Tschirch, the principal constituent of

asafoetida is the ferulic ester of asaresinotannol, which forms about 60 per cent. of the substance. There is present about 7 per cent. of a foul-smelling essential oil.

In judging the quality of asafoetida, the amount of resinous matter soluble in alcohol is of the first importance, as well as the amount of mineral matter, which is, of course, quite inert. The percentage of ash is very variable, sometimes falling as low as 3 per cent. to 5 per cent. in fine, picked tears, and sometimes amounting to 40 or even 50 per cent. in the common lumps. The amount of alcohol-soluble resin will, of course, vary more or less inversely with the amount of mineral matter.

Martin and Moor examined eleven samples of asafoetida, in which they found the ash to vary from 26·4 per cent. to 63·1 per cent., whilst the percentage of resin soluble in alcohol varied from 14·8 to 39·8 per cent. The author has examined many samples of asafoetida of commerce, occurring in lumps, and has found the ash to vary from 6 to 40 per cent., and the resin soluble in alcohol from 30 to 70 per cent.

The United States authorities have recently introduced a so-called "lead number" standard for this gum resin, which is an arbitrary value, based on the amount of lead precipitated from a solution of lead acetate, by a given weight of the resin dissolved in alcohol. The author has shown that this figure is fallacious, and has been supported in this view by other English chemists, so that it need not be further discussed.

MYRRH AND BDELLIUM

The true myrrh of commerce is that known as Herabol myrrh, a gum resin which is derived from various species of *Balsamodendron* and *Commiphora*. Bisabol myrrh is quite different in flavour and odour from Herabol myrrh, and appears to be derived principally from *Balsamea erythrea*.

Myrrh is a fragrant gum resin, varying in colour from pale yellow to almost black. It usually arrives in chests of very mixed qualities, which are either purchased as such ("sorts") or are picked and graded into different qualities. It is found in small masses, from the size of a pea to that of a chestnut, soft when fresh, but becoming hard on keeping. It is used to some extent in medicine, and largely in the manufacture of toilet preparations, perfumery, and incense, as its odour is particularly fragrant. Bisabol myrrh is used in the manufacture of Chinese Joss sticks.

Myrrh, at one time, usually reached this country *viâ* Turkey, but for some years past it has been sent direct to London from India, Arabia and Somaliland. The only adulterants met with, except on rare occasions, are earthy matter and bdellium, or—in the case of Herabol myrrh—the Bisabol variety.

Herabol myrrh may be identified by the following reaction. About ten grains of the powdered gum resin are shaken for about ten minutes with 10 cc. of ether. The liquid is then filtered and 2 cc. of the filtrate are evaporated on a water bath in a porcelain capsule. The capsule containing the dry residue is then inverted over another containing strong nitric acid, so that the residue may come into contact with the fumes. A fine violet coloration results. Bisabol mrryh does not give this coloration.

A good sample of myrrh will contain from 30 to 40 per cent. of resin soluble in alcohol. According to Tucholka, Bisabol myrrh has the following composition—

Gum soluble in water = 22·1%
Gum soluble in alkali = 29·85%

Resin = 21·5%
Ethereal oil = 7·8%
Water = 3·17%
Insoluble vegetable matter = 13·4%

Genuine myrrh only yields about 5 to 8 per cent. to petroleum ether. The following are analytical values obtained by Kremel—

	Soluble in alcohol.	Acid No.	Ester No.
Herabol	39·5%	64	95
Herabol	42%	60·2	116·5
Herabol	23·9%	70·3	145·8
Bisabol	30·7%	42·1	130·8

The essential oil obtained by distilling the gum resin with steam is very aromatic, but rarely seen in commerce.

Bdellium, which resembles myrrh a good deal, is found as African bdellium, and as East Indian bdellium. It is obtained from various species of *Commiphora* and *Balsamodendron*. It comes into commerce in larger pieces than myrrh, and is usually very dirty. Its value lies in its somewhat aromatic properties. Bdellium does not give the violet coloration with nitric acid described above, so that it can be distinguished from myrrh by this means.

Dieterich has examined a number of samples and gives the following results—

	Acid value.	Ester value.
African	12·8	70
,,	14·4	69·3
,,	9·7	96·4
,,	11·9	95·6
,,	19·2	90·7
,,	20·8	90·1
Indian	35·7	46·8
,,	37·2	48·5

AMMONIACUM

This gum resin is produced from an umbelliferous plant, *Dorema ammoniacum*, a plant found principally in Persia. It is used in medicine, and occurs in commerce in the form of lumps or in small tears, the former usually being the smaller fragments which have become agglutinated and thus apt to contain extraneous matter, such as woody fibre or earthy matter. It is hard and brittle, but becomes soft by handling, and possesses a slight, somewhat unpleasant odour, due to the presence of a small quantity of an essential oil.

It is a mixture of gum, resin, etc., in the following proportions—

Gum	.	18 to 28%
Resin	:	50 to 75%
Essential oil	.	1·5 to 6%

The only adulterant usually found present is the so-called African gum ammoniacum, a gum resin derived from *Ferula tingitana*, which may be detected in the following manner. If about sixty grains be powdered and boiled for fifteen minutes with about half an ounce of strong hydrochloric acid, and the liquid cooled and filtered and then rendered alkaline with solution of ammonia, the pure gum resin will show no fluorescence; but if African ammoniacum or gum galbanum be present a blue fluorescence will be found in the liquid. Beckurts and Bruche have recorded the following values for this gum resin—

Specific gravity	=	1·190 to 1·210
Ash	=	0·79% to 447%
Resin	=	56 to 68%
Acid No. of resin	=	69 to 80
Ester No. of resin	=	19 to 38

GALBANUM

This gum resin is fairly closely allied in general character with ammoniacum. It is the product of various species of the umbelliferous plant *Peucedanum*, of which *Peucedanum galbaniflorum* is the chief. It is used to a small extent in medicine, and is found on the market in the form of lumps, and also in small tears. Some samples contain so much essential oil as to be semi-solid in consistence, but the greater part of the supply found in commerce is hard, with a somewhat greasy fracture, and contains about 8 to 10 per cent. of essential oil. In the very soft varieties as much as 22 per cent. has been found. The gum resin contains umbelliferone, together with a number of scarcely well-defined substances of complex constitution. It yields the fluorescent reaction with hydrochloric acid and ammonia described under ammoniacum, which is due to the presence of umbelliferone.

Beckurts and Bruche have recorded the following figures for five samples of the pure gum resin—

	Specific gravity.	Ash.	Resin.	Acid value of Resin.	Ester value of Resin.
(1)	1·110	4·0%	63%	22	82
(2)	1·130	8·7%	56%	19	91
(3)	1·109	4·1%	58%	40	69
(4)	1·133	8·4%	54%	19	63
(5)	1·121	4·9%	60%	25	90

CHAPTER III

BALSAMS, MEDICINAL RESINS, ETC.

GENERALLY speaking, the name balsam is restricted to that class of oleoresinous plant exudations which contain highly aromatic bodies such as benzoic or cinnamic acid, but the application of the name is purely empirical, and as a matter of convenience will not be adhered to. The principal substances known under this description, or for reasons of convenience may be grouped under it, which will be described are the following: balsam of copaiba; balsam of Peru; balsam of tolu; Canada balsam; storax, and gum benzoin.

BALSAM OF COPAIBA

Balsam of copaiba, or capivi, as it used to be called, is an oleoresin obtained from the trunk of the tree *Copaifera landsdorfii*, and other species of *Copaifera* indigenous to South America.

Balsam of copaiba is used to a very large extent in medicine, being one of the few recognised remedies for gonorhoea. It is a mixture of a resin and an essential oil, the proportions varying very greatly according to the locality of production. The ordinary varieties used in medicine, such as Maranham or Maracaibo balsam, contain about 45 per cent. of essential oil and 55 per cent. of resin, whilst the variety known as Para balsam of copaiba contains up to 75 per cent. of resin and 25 per cent. of essential oil.

Balsam of copaiba is a deep brown, viscid liquid, varying in consistency according to the relative proportions of resin and essential oil. The oil is principally composed of sesquiterpenes, and the resin almost entirely of resin acids.

The author has drawn attention to the fact that adulteration of this product is very gross. One of the principal adulterants is the so-called African balsam of copaiba, a product of unknown botanical origin, although almost certainly derived from some species of *Copaifera*. Another very common adulterant is gurjun balsam, the product of various species of *Dipterocarpus*.

Umney and Bennett have recorded the following values for the five following varieties of balsam of copaiba.

	Specific gravity.	Per cent. of oil.	Acid No.	Ester No.
Bahia	0·938	49·7	33·7	15·3
Cartagena	0·970	41·3	56·0	28·0
Maracaibo	0·969	42·5	50·2	12·1
Maranham	0·990	41·8	81·5	12·8
Para	0·920	62·4	33·1	26·9

The resins from the same five samples of balsam had the following characters—

	Acid No.	Ester No.
Bahia	73·1	73·0
Cartagena	135·7	45·1
Maracaibo	80·3	49·9
Maranham	136·3	36·7
Para	68·9	87·2

Essential oil distilled from copaiba balsam has a specific gravity of about 0·895 to 0·908 and an optical rotation from −4° to −35°. The essential oil from African copaiba is dextro-rotatory, and that from gurjun balsam yields the following reaction, which renders it easy of detection. If five drops of the essential oil are dissolved in 5 cc. of glacial acetic acid, to which five drops of strong nitric acid have been added, a pink to purple colour will be developed within a minute.

Adulterations of this valuable balsam with turpentine

and with castor oil used to be practised, but are not met with to-day.

According to M. Utz, a recent new adulterant of copaiba is a cheaper oleoresin, segura balsam, whose botanical origin is not known with certainty. It is a dark brown, viscous thick fluid, with a pleasant aromatic odour; readily soluble in chloroform, benzene, petroleum ether and carbon tetrachloride; partially soluble in alcohol. It yields from 30 to 40 per cent. of essential oil on steam distillation. This oil has the sp. gr. 0·9451 at 15°C. The oleoresin gives a reddish-brown colour with sulphuric acid; yellowish green with nitric acid; red-brown with vanillin and hydrochloric acid; a light brown with chloride of tin in the cold and on heating. Admixture with Maracaibo balsam causes increase of the sp. gr. and the ester value, with a lowering of the acid value; the cold saponification value is only slightly affected. It is probable that but little segura balsam alone is used as the adulterant, but its essential oil, or a mixture of the oil with segura balsam, is employed.

BENZOIN

Benzoin or " gum benjamin," as it is frequently called, is the balsamic resinous matter obtained from various species of *Styrax*, of which *Styrax benzoin* is the principal. The trees yielding this substance are indigenous to Siam, Sumatra and Java. It appears that the trees, when allowed to live normally do not secrete any benzoin, but the infliction of a wound on the tree, sufficiently deep to injure the cambium layers of cells results in the formation of numerous oleoresin ducts, in which the secretion at once commences to make its appearance. Gum benzoin may, therefore, be described as a pathological secretion. The trunks of the styrax trees are usually hacked to the proper depth by means of an axe, and the secretion soon commences to accumulate beneath the bark, or to exude from the incisions. When the secretion has sufficiently hardened it is collected and packed in boxes for export. The following varieties of gum benzoin are commercial articles.

Siam benzoin is one of the most esteemed varieties, if not the most esteemed, of benzoin. It is chiefly collected in the province of Luang Pratang, but the exact species which yields it is still a matter of uncertainty. It is found either in lumps or in comparatively large tears, the latter being practically free from extraneous matter, and therefore of a higher degree of purity. Siam benzoin is characterised by its pronounced odour of vanillas, and by its freedom from cinnamic acid.

Sumatra benzoin is undoubtedly the product of *Styrax benzoin*, and is produced on the Island of Sumatra. It always occurs in block form, which consists of white tears agglutinated by a semi-vitreous resin of dull red or greenish colour. The odour of this variety recalls that of storax rather than vanilla, and it contains cinnamic acid, which distinguishes it from the Siam variety. This may be detected by boiling the specimen with dilute sulphuric acid and potassium permanganate when a marked odour of almonds, due to the formation of benzaldehyde, results, which is not the case with Siam benzoin.

Palembang benzoin is another Sumatra type of benzoin, derived from trees, whose botanical origin is not known. It consists of lumps in which only a few tears are to be found, and has only a slight odour, which resembles, to a weak degree, that of ordinary Sumatra benzoin. It is the cheapest of all benzoins, and is chiefly used for the manufacture of benzoic acid.

Gum benzoin is used to a fair extent in medicine. It

is a constituent of the well-known substance, "Friar's Balsam," and is also used largely for the manufacture of benzoic acid, although the cheapness of the synthetic product in normal times has militated against any large employment of the natural acid. It is also used in perfumery, and in the manufacture of incense, fumigating pastilles and similar preparations.

The principal constituent of Siam benzoin is the aromatic acid, benzoic acid, partly free and partly combined with resin alcohols, which have been named benzoresinol and siaresinotannol. Vanillin is present in traces, and, in Sumatra benzoin, about half the free acids consist of cinnamic acid. Small quantities of vanillin, styrol, styracine, phenyl-propyl cinnamate and benzaldehyde, are also present in Sumatra benzoin.

Benzoin is liable to be adulterated with a considerable amount of vegetable debris and earthy matter. In good, natural, specimens, the total amount of matter insoluble in 90 per cent. alcohol should not exceed 10 per cent., and the ash should not be more than 2 per cent. or at most 2·5 per cent. The author has examined a number of samples which were all of excellent quality and which gave the following results—

Siam Benzoin.

Mineral matter.	0·24 to 1·98%
Soluble in 90% alcohol	88% to 96·4%
Acid value	130 to 158
Ester value	42 to 69

Sumatra Benzoin.

Mineral matter.	0·4 to 1·96%
Soluble in 90% alcohol	90% to 93·5%
Acid value	98 to 139
Ester value	51 to 98

Other varieties.

Mineral matter.	0·4 to 2·85%
Soluble in 90% alcohol	86% to 95%
Acid value	106 to 142
Ester value	51 to 91

Good qualities of benzoin contain from 12 to 20 per cent. of benzoic acid, which may be determined approximately by powdering the sample, mixing it with twice its weight of sand and heating it in a beaker covered with a perforated filter paper. The benzoic acid sublimes and is condensed in a porcelain cone kept cold by a current of cold water.

Benzoic acid and its sodium salt, sodium benzoate are both largely used as preservatives, being generally preferred to salicylic acid as there is a prejudice against the latter (probably unjustified) on account of the belief that there may be undesirable after effects following its use.

BALSAM OF TOLU

Balsam of Tolu is the product of the trunk of the tree *Myroxylon Toluifera*, a tree indigenous to New Granada. It is an aromatic balsam exuding from artificially-made incisions in the trunk of the tree, and is collected in gourds, and finally exported in tins. When fresh, balsam of tolu is a soft, tenacious, resinous substance becoming much harder on keeping, and quite brittle in cold weather. If a small fragment be warmed and pressed between two pieces of glass, crystals of cinnamic acid can be discerned throughout the mass, when examined under the microscope.

Balsam of Tolu has a fragrant odour and an aromatic and slightly acid taste. It is used in perfumery and in the preparation of various aromatic substances, and also in medicine.

It contains benzyl benzoate, benzyl cinnamate, free cinnamic and benzoic acids, traces of vanillin and the benzoic and cinnamic acid esters of toluresinotannol, an alcohol of very complex constitution.

The principal adulterants of balsam of tolu are

ordinary colophony and Balsam of Tolu which has previously been deprived of most of its aromatic substances by a process of extraction. A genuine balsam will yield not less than 90 per cent. to alcohol, 95 per cent. to chloroform, and only 3 to 8 per cent. to petroleum ether. A normal sample will have the following characters—

Acid value = 105 to 140
Ester value = 38 to 70

If 5 grammes are warmed with two successive portions of 25 cc. and 10 cc. of carbon bisulphide, the solutions, when evaporated should yield a distinctly crystalline residue, which should require not less than one-third of its weight of caustic potash for saponification.

Colophony may be detected as follows: Five grammes are exhausted with petroleum ether, and the filtered petroleum ether extract is shaken with an equal volume of a 0·1 per cent. aqueous solution of acetate of copper. If colophony be present, the petroleum ether will be coloured green, as the abietic acid present in colophony forms a green copper salt soluble in petroleum ether.

Balsam of Tolu is official in the British Pharmacopoeia, which requires it to have the following characters.

A soft, tenacious solid when first imported, becoming harder and finally brittle. Transparent and yellowish-brown in thin films. Pressed between pieces of glass with the aid of heat, and examined with a lens, it exhibits crystals of cinnamic acid. Odour fragrant, especially when warmed; taste aromatic and slightly acid. Soluble in alcohol (90 per cent.) the solution being acid to litmus. Acid value, 107·4 to 147·2; saponification value, 170 to 202. If 5 grammes are gently warmed with three successive portions of 25, 15 and 10 millilitres of carbon disulphide, the solution yields, when evaporated to dryness, a distinctly crystalline residue, which when

MYROXYLON PEREIRAE
(Peru Balsam Tree)

tested as described under "Styrax Preparatus" yields not less than 1·25 grammes of balsamic acids.

BALSAM OF PERU

Balsam of Peru is an oleoresinous liquid obtained from the bark of *Myroxylon pereirae*, a tree found to a comparatively large extent in the forests of San Salvador in Central America. The secretion is not a normal one, and does not occur naturally in the tree. It may be described as a pathological secretion, induced by gently heating the bark, and subsequently scorching it. The wounds made in the tree are covered with rags which are continually changed, and pressed and boiled with water, the purified resin being then packed in tins and exported, principally from Acajutla and Belize to New York, and prior to the war to Hamburg, whence it reaches the London market.

Balsam of Peru is a viscid, but not glutinous, reddish-brown liquid of highly aromatic odour. It is used in medicine, and to a certain extent in perfumery. It is used to a considerable extent in the treatment of skin diseases, being a powerful antiseptic and parasiticide. It was formerly used in cases of phthisis, but its value in this direction is now recognised to be nil, and it is not now so employed.

In commerce the only variety of Balsam of Peru of any importance is the dark brown type above indicated. There exists a pale-coloured variety, known as white Balsam of Peru, apparently derived from the fruit of the same tree, but this is now but rarely met with in commerce. An inferior grade of balsam is prepared by boiling the bark, and is known as Balsam of Cascara or Tacuasonte, but this is not now met with to any extent.

Balsam of Peru is adulterated to a considerable extent, as it is a somewhat expensive product. Fatty oils, such as castor oil and olive oil are favourite adulterants, and colophony and gurjun balsam have been met with. To-day, however, the most difficult adulterant to deal with is the so-called synthetic Balsam of Peru, a mixture of aromatic substances which very closely resembles the natural balsam in appearance and odour.

The balsam has been the subject of numerous investigations, but unfortunately a number of these are of little value, as the specimens examined have been imported from the producing district, without due care having been taken as to their authenticity, and were almost certainly adulterated.

Pure Balsam of Peru contains from 55 to 65 per cent. of cinnamein, by which term is meant a mixture of esters of cinnamic and benzoic acids, a certain amount of free aromatic acids and vanillin, and there is also a small amount of free alcoholic bodies. The alcoholic constituents of the esters appear to consist mainly of benzyl alcohol and peruviol.

Five samples of apparent authenticity have been examined by Dr. Preuss, and were found to have the following characters—

	Specific gravity.	Cinnamein.	Ester No. of Cinnamein.	Resin.
(1)	—	61%	240	20%
(2)	1·1404	64·7%	260	18·1%
(3)	1·1408	66%	260	16·8%
(4)	1·1612	50·8%	249·8	28·4%
(5)	—	37·7%	—	27·55%

Sample No. 5 was not a normally prepared sample, but one which was made by the extraction of the bark with ether.

Dieterich has investigated this valuable product very fully, and has given the following summary of his results (*Analysis of Resins:* K. Dieterich; Scott,

Greenwood & Co.), for three samples of absolutely authentic balsams, obtained from Honduras—

	(1)	(2)	(3)
Acid value	77·46	76·92	27·34
Ester value	165·61	137·42	137·67
Saponification value	243·07	214·34	215·01
Cinnamein, etc.	71·41%	77·56%	73·63%
Resin esters	15·7%	13·18%	17·32%
Insoluble in ether	4·38%	4·31%	3·57%

The refractive index of pure Balsam of Peru lies between 1·4800 and 1·4855 at 20°.

High acid values indicate the presence of colophony as an adulterant, which will also cause the ester value to fall, and will also yield the copper acetate test with the petroleum ether extract, described under Shellac and Balsam of Tolu.

Tolu balsam is occasionally used as an adulterant of Balsam of Peru. This will cause the acid value to rise and the ester value to fall. Copaiba balsam increases the amount of essential oil obtainable on distillation, since genuine Peruvian balsam only contains a minute quantity of essential oil, whereas balsam of copaiba contains from 40 to 75 per cent. The ester number will also be materially diminished.

Storax, another adulterant sometimes met with, reduces the ester value. Gurjun balsam may be detected by the high yield of essential oil which can be obtained on distillation, in which respect it is similar to copaiba balsam. It may be differentiated from the latter by the purple coloration given when five drops of the essential oil are mixed with 10 cc. of glacial acetic acid containing five drops of nitiric acid.

Aguilar, in a recent publication, recommends the following tests for Balsam of Peru. When treated with twice its volume of petroleum ether, the balsam gives a *clear* liquid floating on the surface of the undissolved portion.

When treated with twice its volume of sulphuric acid, no sulphur dioxide should be evolved if the balsam is pure.

If mixed with half its weight of slaked lime, and heated in the water bath for half an hour, pure Balsam of Peru should not solidify.

To detect artificial Balsam of Peru, an ethereal solution should be gently poured on to the surface of a small quantity of strong sulphuric acid, and a little hydrochloric acid then cautiously added. In the presence of artificial balsam, there appears a lower ring at the junction of the liquid, of a red colour and an upper ring of green colour, which do not appear in the case of the natural balsam.

Balsam of Peru is official in the British Pharmacopoeia, which requires it to have the following characters—

A viscid liquid, in bulk nearly black but in thin layers deep orange-brown or reddish-brown and transparent. Agreeable balsamic odour; taste acrid, leaving when swallowed a burning sensation in the throat. Insoluble in water; soluble in chloroform. One volume is soluble in one volume of alcohol (90 per cent.), but on the further addition of two or more volumes of the alcohol the mixture becomes turbid. Specific gravity between 1·140 and 1·158. Does not diminish in volume when shaken with an equal bulk of water (absence of ethylic alcohol). When tested by the following method it yields not less than 57 per cent. of cinnamein, the saponification value of which is not less than 235.

Dissolve 1 gramme of the balsam in 30 millilitres of ether and shake in a separating funnel with two successive quantities of 20 and 10 millilitres of N/2 solution of sodium hydroxide. Separate the alkaline solutions, mix and shake with 10 millilitres of ether. Draw off and reject the alkaline solution. Add the second ethereal solution to that previously obtained. Wash the mixed ethereal solutions with two successive quantities of 5 millilitres of water. Transfer the ethereal solution thus washed to a tared wide-mouthed flask, evaporate at a gentle heat until the odour of ether has disappeared, add 1 millilitre of absolute alcohol, dry at 100° for half an hour, and weigh. The weight of the cinnamein thus obtained is not less than 0·57 gramme. To this residue add 20 millilitres of N/2 alcoholic solution of potassium hydroxide and 20 millilitres of alcohol (90 per cent.). Attach a reflux condenser, boil for half an hour, and titrate back again with N/2 solution of sulphuric acid, solution of phenolphthalein being used as indicator. Each gramme of the residue thus treated requires not less than 8·4 millilitres of the alkaline solution for complete saponification (corresponding to a saponification value of not less than 235).

Storax

Storax or Styrax is a liquid oleoresinous substance of very considerable commercial value. It is the product of the trunk of the tree *Liquidambar orientalis*, one of the family of the *Hamamelideae*, which is indigenous to the south-western portion of Asiatic Turkey. As is the case with the nearly related products, Balsam of Tolu and Balsam of Peru, storax is not a normal secretion of the plant tissues, but is to be regarded as a pathological product, the formation of which has to be induced by beating the bark of the tree. The object of this beating is to inflict an injury on the cambium layers, when numerous oleoresin ducts are formed, in which the oleoresinous material is secreted, and whence it is discharged into the wounded bark. The bark is peeled off, pressed, and boiled with water and again pressed, and the oily layer separated from the aqueous

INDIAN PINE FOREST ON ROCKS

From "Indian Forest Memoirs" (*R. S. Troup*)

liquid. The crude storax thus obtained is an opaque greyish liquid of very thick consistency, and having a sweet, aromatic odour. In this condition it usually contains about 20 to 30 per cent. of water, together with fragments of bark and other vegetable debris. On standing the water separates from the oleoresin, and the latter is purified by solution in alcohol, filtering the alcoholic liquid and evaporating the solvent, the residual oleoresinous matter, being known as purified storax. This is a dark brown viscous liquid of powerful aromatic odour and taste.

Prepared or purified storax is used as a drug and is official in the British Pharmacopoeia which requires it to have the following characters—

Brownish-yellow, viscous, transparent in thin layers. Entirely soluble in alcohol (90 per cent.) and in ether. Odour and taste agreeable and balsamic. Boiled with solution of potassium chromate and sulphuric acid it evolves an odour of benzaldehyde. Loses not more than 5 per cent. of its weight when heated in a thin layer on a water-bath for one hour. Acid value not less than 60 and not more than 90; ester value not less than 100 or more than 146. Yields not less than 20 per cent. by weight of cinnamic acid when tested by the following process—

Dissolve 2·5 grammes of the storax in 25 millilitres of N/2 alcoholic solution of potassium hydroxide, boil for one hour under a reflux condenser, neutralise with N/2 solution of sulphuric acid, remove the alcohol by evaporation, and dissolve the residue in 50 millilitres of water. Shake this aqueous solution with 20 millilitres of ether; after separation remove the ethereal layer, wash it with 5 millilitres of water and add the washings to the aqueous solution, rejecting the ethereal liquid. Acidify the aqueous solution with diluted sulphuric acid and shake it with four successive portions, each of 20 millilitres, of ether. Mix the ethereal solutions, wash with a few millilitres of water, transfer to a flask and distil off the ether. To the residue add 100 millilitres of water and boil vigorously for fifteen minutes under a reflux condenser. Filter the solution while hot, cool to 15·5°, and collect on a tared filter the crystals of cinnamic acid that have separated. Repeat the extraction of the residue with the filtrate at least three times, or until no more cinnamic acid is removed. Press the filter paper and crystals between blotting-paper, dry in a desiccator over sulphuric acid, and weigh. Add to the weight of the crystals so ascertained 0·03 gramme (representing the average amount of cinnamic acid remaining dissolved in the aqueous liquid). The total weight is not less than 0·5 gramme.

There is a product on the market known as storax calamita, which, however, has nothing to do with the ordinary " gum storax " of commerce. It is the powdered bark of the American tree *Liquidambar styraciflua* from which as much resinous matter as possible has first been extracted.

Storax is adulterated with fatty oils, but the present practice is not to add an adulterant, but to extract as much as possible of the odoriferous constituents, and sell the emasculated storax as genuine.

Dieterich gives the following figures as representative of genuine storax—

Acid value of alcoholic extract	37·2 to 96·6
Ester value ,, ,,	74·6 to 168
Saponification value of extract .	134·6 to 249
Ash	0·07% to 1·20
Volatile at 100°	10·25 to 40·15%
Soluble in alcohol . .	56 to 84%
,, acetic ether .	69 to 73·6%
,, chloroform .	69 to 72·6%
,, ether	65·8 to 82·8%

Soluble in benzene . 64·8 to 74·8%
,, turpentine . 54·4 to 57·8%
,, carbon sulphide . 62·3 to 67·8%
,, petroleum ether . 15·0 to 19·4%

The specific gravity of storax varies between 1·109 and 1·125. The figures given above are not necessarily exhaustive, and it is probable that very few genuine samples give results outside much narrower limits.

At the present time, by far the greater number of samples of storax found on the market are merely the residues left after the extraction of the valuable odoriferous constituents, as far as can be achieved, and they therefore have but slight medicinal or perfume value.

This question is one of great importance to users of storax, and has received a good deal of attention during the past few years from analysts. J. C. Umney has persistently called attention to the quality of this balsam, and in an article in the *Perfumery and Essential Oil Record* (1911, 126), states that in the previous four years, the characters of the storax to be found in the London market had so radically altered, that it became necessary to inquire whether there had been any alteration in the botanical origin of the balsam, or any modification in the method of its preparation, or whether adulteration—either by addition or abstraction was the real cause of the alteration in character. As a result of careful inquiries, Umney was able to decide that no alteration in the botanical source of supply has taken place, and no deviation from the usual methods of production, and that when the storax is imported direct from the place of production—the forests of the southwest of Asia Minor—it possesses its old-time fragrance, and its usual high value for medicinal and perfumery purposes. But the shipments from certain ports, notably Trieste and Marseilles, show that the character of the balsam imported through second-hand sources

Photo by   *Henry Irving.*

LIQUIDAMBAR STYRACIFLUA

has altered materially, on account of the abstraction of a large amount of the essential constituents, which cause the so exhausted balsam to have very little value for either medicinal or perfumery purposes. The extraction of the valuable portion of the storax appears to serve two purposes. One of these is for the manufacture of a concentrated essence of storax prepared by extraction with a suitable solvent from the storax, and stated to be about fifty times as strong as natural storax in odour value. The other, and more important purpose, is the abstraction of the cinnamic alcohol, which possesses a sweet, hyacinth-like odour, and is used as a very serviceable fixative, especially in soap perfumery. The ordinary synthetic cinnamic alcohol is not nearly so sweet as the product obtained from storax, and so long as the impoverished balsam can be sold, it is obvious that the cinnamic alcohol will continue to be abstracted. Two old samples of authentic origin were examined by Umney, and found to have the following characters—

	Acid No.	Ester No.	Cinnamic acid.
(1) Crude	50.6	100.4	20.6%
(2) Strained	55.2	126.6	26.3%
(3) Purified from (1)	60.1	130.1	25.5%

In the following table are shown the characters of a number of samples examined by Umney between the years 1907 and 1911 in which the change referred to can be seen.

Year.	Acid No.	Ester No.	Cinnamic Acid, Free and Combined.
1907	68.9	111.9	19.0
1907	67.1	120.9	15.2
1908	96.4	94.0	14.1
1909	95.9	65.4	11.6
1909	111.6	63.8	11.7
1910	101.5	92.4	9.3
1910	93.7	84.4	8.2

Year.	Acid No.	Ester No	Cinnamic Acid, Free and Combined.
1910	97·1	90·2	8·3
1910	99·4	30·3	7·6
1911	110·3	82·8	7·5
1911	107·0	81·1	5·6
1911	99·7	14·5	4·0
1911	100·1	79·6	3·5
1911	96·5	72·4	2·5

In view of this wave of adulteration that has set in for storax, Umney has recommended the following as the standard tests which should hold good for the balsam.

Acid and ester values.—Dissolve 2·5 grms. in 20 cc. of alcohol, add a few drops of phenol-phthalein solution and titrate with semi-normal alcoholic potash until a permanent pink coloration is produced. Not less than 5·3 and not more than 8·0 cc. should be required. (Acid No. 60 to 90.) Add to this solution 20 cc. of semi-normal alcoholic potash and heat to boiling for an hour. Titrate back the excess of potash by means of semi-normal sulphuric acid. Not less than 9 cc. and not more than 12·6 cc. of semi-normal potash should be absorbed. (Ester No. 100 to 140.)

Total Cinnamic Acid (Free and Combined).—Evaporate the alcohol from the saponified solution, and dissolve the residue in 50 cc. of water. Transfer to a separator, wash with 10 cc. of ether and reject the ethereal layer. Acidulate the aqueous solution with normal sulphuric acid and extract the liberated acids with ether. Evaporate the ethereal solution and extract the residue with 100 cc. of boiling distilled water. Filter while hot, allow to cool to 15° and collect the crystals of cinnamic acid on a counterbalanced filter. Twice repeat the extraction of the residue with the filtrate heated to boiling and collect the crystals.

Dry the crystals at 100°C. and weigh. Add ·030 grms. to correct for solubility. At least 0·375 grms. should be obtained from 2·5 grms. corresponding to at least 15 per cent. of free and combined cinnamic acid. The crystals obtained should give the reactions of cinnamic acid. Examined by this method the best samples yield 20 to 25 per cent. of total cinnamic acid.

Canada Balsam

Canada balsam, or Canada turpentine, is the oleoresinous exudation of the balsam fir, a coniferous tree indigenous to the United States and Canada. The oleoresin, which consists of a resinous substance dissolved in an essential oil, is secreted in schizogenous vessels in the bark, and collects in small cavities, which eventually become blisters. These are punctured and the oleoresin collected.

Canada balsam forms a viscid yellow liquid, frequently showing a distinct green fluorescence. It has a pleasant, turpentiny odour and a bitter acrid taste. On keeping, the essential oil gradually evaporates and the resin dries to a hard, transparent varnish.

This oleoresin is used to a certain extent in medicine, and is a constituent of flexible collodion, and is of great service in the mounting of microscopic specimens. In the last-named purpose it is dissolved in an equal volume of xylene. Such a solution forms a non-crystallising mounting medium, having a refractive index approximating to that of ordinary glass, and therefore involving the minimum dispersion of light. A genuine Canada balsam has the following characters—

Specific gravity	0·985 to 0·995
Optical rotation	$+1°$ to $4°$
Refractive index	1·5180 to 1·5210
Acid value	84 to 87
Ester value	5 to 10

RESINS, ETC.

On distillation, Canada balsam yields from 15 to 25 per cent. of essential oil which boils at about 160°, is laevo-rotatory, and consists almost entirely of terpenes. Canada balsam oil does not differ materially in composition or properties from ordinary turpentine oil.

The commonest adulterant of Canada balsam is, as above mentioned, either colophony or crude turpentine. The best method of detecting colophony is to drive off the essential oil by prolonged heating on a water bath, leaving the dry resin to be examined. In the case of pure Canada balsam, this dry resin will be found to have an acid value of about 120 to 125, whilst colophony has an acid value of about 165, so that an acid value of over 130 is a very strong indication of the presence of added colophony or crude turpentine. The presence of these adulterants can, according to E. Dieterich, be confirmed by the Storch-Morawski reaction, but, in the author's opinion, this is not correct and cannot be relied on.

There is a product sometimes to be found on the market under the name of Oregon balsam, which is not a natural product at all, but is a mixture of common rosin and turpentine, closely resembling Canada balsam in appearance, and which is used to adulterate it frequently. There is also a true Oregon balsam, which is the oleoresinous exudation from the trunk of *Pseudotsuga mucronata*. This balsam yields a considerable quantity of an essential oil, which can be characterised by its consisting very largely of laevopinene. The factitious oregon balsam of the above description contains the same proportion of solid resin as does pure Canada balsam, and the physical characters of the two substances are as nearly as possible the same. Detection of such adulteration is however possible, by separating the essential oil and the resin and examining each separately.

BURGUNDY PITCH

This resin, or oleoresin, is used to a small extent in medicine, especially in the manufacture of plasters. It is the resinous exudation from the stems of *Picea excelsa*, one of the coniferous family, and is principally obtained in Finland and the Black Forest. Incisions are made in the bark of the tree and the oleoresinous matter which exudes is scraped out of the wounds in the trunk, melted under water and then strained. It is an opaque, brittle mass, yellow or yellowish-red in colour, but sufficiently soft to gradually assume the shape of the vessel in which it is stored. ' It breaks with a clean, conchoidal fracture, and has an aromatic odour and sweetish aromatic taste.

An imitation Burgundy pitch is made by melting together common pitch, rosin and a little turpentine, and mixtures of this type constitute the greater part of the Burgundy pitch of commerce. The pure resin is soluble in twice its weight of glacial acetic acid, whereas most samples of the factitious resin are not clearly soluble.

The principal constituents of Burgundy pitch are several complex resin acids, with small quantities of esters and neutral resins, with a small amount of essential oil.

The acid value of Burgundy pitch varies from 135 to 148.

VENICE TURPENTINE

Venice turpentine, or larch turpentine, is the oleoresinous exudation of *Larix Europaea*, or *Larix decidua*. It is a viscous liquid of a yellowish or yellowish-green colour, and having a characteristic pleasant smell and somewhat bitter taste, the latter feature probably being due to the presence of a trace of a bitter glucosidal principle. The oleoresin itself is dextro-rotatory, whilst the essential oil is laevo-rotatory.

Venice turpentine is employed to a small extent in medicine, but is rarely found in a pure state. Most of the commercial oleoresin is a factitious article, made by mixing rosin, rosin oil and oil of turpentine together. This so-called Venice turpentine is used to a limited extent in the manufacture of sealing-wax and certain types of varnish, and also, improperly, as a substitute for the genuine article, in medicine.

The following are the analytical figures of the genuine oleoresin—

Acid value	65 to 75
Ester value	30 to 55
Saponification value	95 to 128
Iodine value	145 to 155

If the sample is factitious a mixture of the type described above, it will be indicated by the following alteration in the above figures. The acid value will be considerably raised and the ester value correspondingly lowered. The iodine value will usually be lowered. If rosin oil has been used, as is generally the case, there will be a considerable amount of unsaponifiable matter present, which can be extracted by means of ether from the aqueous saponification residue. Fabris recommends the following test for rosin oil. Dissolve 5 grammes in 20 cc. of 95 per cent. alcohol and add a 10 per cent. solution of caustic potash till the liquid is alkaline. Pure Venice turpentine remains clear, whilst, in the presence of rosin oil, oily drops separate out.

GARCINIA RESIN

Garcinia resin is the sap of an unknown species of the tree, obtained from the Federated Malay States. It belongs to the somewhat restricted class of oleoresinous materials, known as "natural varnish," of which the best known type is that collected in China and Japan from *Rhus vernicifera*, the so-called varnish tree. A similar, but less known product is obtained in Burma from *Melanorrhoea usitata*.

These natural varnishes, as they are called, are obtained from their respective trees as almost colourless viscous fluids, which, on exposure in thin films to a moist atmosphere, dry to hard, nearly black lustrous surfaces, similar to those so familiar to us as the surfaces of Oriental lacquer ware. They are used almost exclusively in their country of origin, but there is no reason why the exportation of these products should not be encouraged, as they would probably find considerable employment in decorative work in this country. A sample has been examined in the laboratories of the Imperial Institute, which was received from the Curator of the Perak Museum. It was described as the sap of a species of garcinia, obtained by making incisions in the bark of the trees. When collected it is boiled until it is as thick as ordinary cream, a little turpentine is added, and it is then ready to be used as varnish, being applied by means of a pad of cloth. In its prepared condition it forms a pale yellow emulsion, and if boiled till all the water is evaporated it solidifies and cannot again be dissolved in turpentine, but if water be added at once it can again be made into an emulsion. The watery emulsion will only keep for a few days, as fermentation sets in, hence the need of using turpentine in its finished form. This Malay varnish is claimed to be quite equal in whiteness and hardness to the best Japanese lacquer. The tree which yields it is a wild one, and as its fruits develop freely, there should be no difficulty in planting it.

The resin itself is soluble in the usual organic solvents, such as turpentine, chloroform, benzene and ether, and almost completely so in alcohol. It melts at about 65°C., and leaves only a trace of ash on ignition. Its

RESINS, ETC. 63

acid value is 89·2 and ester value 3·3. It therefore consists largely of resin acids, with only a small amount of esters. When dissolved in turpentine oil and used as a varnish, it dries with a coat similar to that produced by good dammar varnishes.

ILLURIN BALSAM

Illurin balsam is an oleoresin closely simulating balsam of copaiba in its general characters. It is known also as West African copaiba. It has been stated to be the product of a tree called *Hardwickia manii*, but later researches have rendered it more probable that it is derived from *Daniella thurifera*. The oleoresin is used by the African natives as a substitute for the true balsam of copaiba.

Illurin balsam is a viscid, yellowish-brown liquid having the following characters—

Specific gravity	0·985 to 0·995
Acid value	55 to 60
Ester value	6 to 10

It yields about 55 to 60 per cent. of resin, which has characters very similar in all respects to the resin from ordinary copaiba.

This oleoresin is used to some extent locally as a remedy for venereal disease, and is employed as an adulterant of genuine copaiba. The essential oil is also used to some extent as an adulterant of more expensive oils.

GURJUN BALSAM

Gurjun balsam is an oleoresin, also very similar to copaiba balsam, which is obtained from various species of *Dipterocarpus*. It is sometimes, but improperly, described as East Indian balsam of copaiba, and is also known as " wood oil."

It is a thick, brownish liquid having the characters shown on the next page.

Specific gravity . . 0·960 to 0·985
Acid value . . . 10 to 25
Ester value . . . 1 to 10

It contains from 40 to 50 per cent. of resin, which differs from copaiba resin by containing far less free acids. It contains a neutral resene, and probably a crystalline alcohol, which has been named gurjunol.

The presence of gurjun balsam as an adulterant can be detected by the following test. The essential oil is distilled off, and about 5 drops added to 10 cc. of glacial acetic acid mixed previously with five drops of nitric acid. If gurjun balsam be present, a reddish-purple to violet coloration will develop within a minute.

PODOPHYLLIN RESIN

The resins of two species of podophyllum are known, and used in medicine as purgatives. These are *Podophyllum peltatum*, the American variety, and *Podophyllum emodi*, the Indian variety. The American drug yields about 5 to 6 per cent. of resin and the Indian drug 9 to 12 per cent.

The resin is prepared by extraction by means of 95 per cent. alcohol, and contains podophyllotoxin, together with an indefinite amorphous resin termed podophyllo-resin. According to Evans (*Analytical Notes*, 1911, 6, 58), three samples of the resin showed the following characters on examination—

RESIN GUM.

	P. peltatum.	P. peltatum.	P. emodi.
Soluble in 90% alcohol	99·8%	99·5%	—
Insoluble in 7% ammonia	·0·8%	10·0%	—
Ash	—	0·5%	—
Acid value	174	145·6	110·1
Ester value	95·5	119·0	95·8
Saponification value	269·5	264·6	205·9
Iodine value	55·6	57·6	44·2

Umney has found that the Indian resin contains as much as 50 per cent of podophyllotoxin, which is one of the principal active constituents of the resin (Schofield reports as much as 63 per cent.), as against about 23 per cent. in the resin prepared from American rhizomes. It is, however, a matter of some uncertainty as to the exact nature of the constituents of the resin to which it owes its physiological activity, and further researches are required in this direction. Until this knowledge of the constituents is forthcoming, it is obvious that any analytical method of standardisation is of doubtful value.

SCAMMONY RESIN

Scammony, or virgin scammony, as it is usually called, is a gum resin obtained by the incision of the living roots of *Convolvulus scammonia*, a plant of the natural order Convolvulaceae. In collecting the resinous material, the root is cut off obliquely, and the emulsion which exudes is collected in a shell at the lower end of the cut surface. The resinous exudation is made into cakes and allowed to dry.

The finest scammony of commerce, which is the type known as virgin scammony, is found in large, flat greyish-black pieces, or irregular flattened lumps, which are easily fractured. The fractured surface is glossy and usually exhibits small cavities, which appear to be produced by a fermentative change which occurs during the slow drying of the gum resin. The odour recalls that of cheese, and the taste is slightly acrid. If the gum resin be very rapidly collected and dried, the colour may be golden-yellow, and the above-mentioned cavities are, of course, absent.

Scammony is frequently adulterated, the principal adulterants being chalk, or other forms of earthy matter, and starch. The ash of the very finest specimens of the gum resin may fall to about 3 per cent., but good commercial samples will usually contain from 6 to 12 per cent. of mineral matter. Traces of starch natural to the drug are often present, but foreign starch, deliberately added, is easily detected by a microscopic examination. Adulteration with foreign resins may be detected by dissolving the resin extracted with ether, in a hot solution of caustic alkali, and then acidifying. Pure scammony resin is not precipitated by this treatment, whereas practically all possible adulterating resins are thrown down. There is a mixture of farinaceous matter with a certain proportion of the genuine gum resin known in commerce under the name skilleep.

The gum resin rarely contains less than 70 per cent. of true resin, unless it has been deliberately tampered with. The Aleppo scammony of commerce is generally the most grossly adulterated. The resin—which is a drug employed in medicine as a powerful purgative, is prepared by exhausting the root of the plant with alcohol, when, on evaporation of most of the solvent, and precipitation of the solution with water, the resin is thrown down. Genuine scammony resin has the following characters.

Acid value	. .	14 to 21
Ester value	. .	200 to 225
Iodine value	. .	10 to 15

There is a Mexican plant, *Ipomoea orizabensis*, which yields a resin very similar in every respect to the scammony resin. This product is known as Mexican scammony resin. It has been examined by Taylor, who gives the following values for a number of samples of true and Mexican scammony resin.

	Acid value.	Ester value.	Iodine No.
True resin	21·1	211·3	13·3
,,	15·5	222·5	10·8
,,	15·6	219·8	13·0
,,	18·2	221·7	14·3
,,	18·8	218·1	14·6
Mexican resin	15·5	171·1	8·7
,,	21·5	165·6	11·5

The optical rotation of scammony resin is characteristic. According to Guignes, the specific rotation of the true resin in alcoholic solution is $-18°30'$ to $-24°30'$, whilst that of colophony is $+ 6°$ to $+ 7°$, of mastic $+ 29°30'$, and of sandarac $+ 31°$ to $+ 34°$.

The only use for this gum resin and resin is in medicine, where it finds a fairly extensive employment.

The most exhaustive recent analyses of scammony resin are those of Engelhardt & Schmidt (*Proceedings of the American Pharmaceutical Association*, 1910, 58, 1027). The following eight samples had, according to all probability, the origins indicated—

1. Genuine scammony from *Convolvulus scammonia*.

2. The same, but purified by alcohol, and so pure scammony resin.

3. Pure scammony resin, obtained by the authors from the roots.

4. Pure resin from roots of *Ipomoea orizabensis*.

5 and 6. Labelled true scammony resin, but probably only scammony gum resin.

7. Mexican scammony.

8. The same, purified by alcohol treatment.

TABLES I. CHARACTERS.

Sample.	Moisture.	Ash.	Acid No.	Sap. No.	Ester No.
I	6·16	2·70	18·5	207·2	188·7
II	1·95		10·6	236·6	226·0
III	2·07	0·21	16·3	256·2	239·9
IV	1·45	0·20	10·2	175·8	165·6
V	2·25	0·07	12·2	177·1	164·9
VI	2·23	0·20	14·0	171·6	157·6
VII	4·29	0·30	13·6	183·8	170·2
VIII	2·03	0·15	14·9	175·9	161·0

TABLE II. SOLUBILITIES.

Sample.	Soluble in Abs. Ether.	Soluble in U.S.P. Ether.	Soluble in $CHCl_3$	Soluble in Alcohol.
I	71·8	85·0	82·1	90·6
II	100·0	100·0	100·0	100·0
III	93·9	96·0	100·0	100·0
IV	89·4	84·1	98·0	100·0
V	90·2	85·5	98·9	100·0
VI	88·3	80·9	96·1	100·0
VII	89·6	82·0	96·9	100·0
VIII	90·4	91·5	97·4	100·0

TABLE III.

Sample.	Iodine No.	Specific Rotation, Degrees.
I	11·69	−25·98
II	10·45	−24·97
III	17·83	−24·24
IV	11·60	−32·78
V	11·48	−33·80
VI	13·93	−34·27
VII	12·46	−31·31
VIII	11·65	−31·83

JALAP RESIN

Jalap resin is prepared from the roots of *Ipomoea purga* by exhaustion with alcohol, driving off most of the solvent, and adding water to the residual solution, by which means the resin is precipitated. This resin consists of two glucosidal resins, of which about 90 per cent. is jalapin and 10 per cent. scammonin. These may be more or less separated by treatment with ether, which leaves the jalapin undissolved, and it is this substance which is sold under the name of Jalapin. Its sole use is as a purgative, for which purpose it is employed to a very considerable extent in medicine. It forms a constituent of many of the pills advertised to the public. It is probable that jalapin is not a chemical individual, but a complicated mixture of bodies,

The roots of the drug contain from 5 to 12 per cent. of resinous material, and the author has examined four samples of the purified resin, with the following results—

Acid value.	Ester value.
14·6	116
13·0	124
15·0	120
16·5	122

According to Dieterich adulteration of jalap resin affects the analytical values in the following manner—

	Acid value.	Saponification value.
Pure Jalap resin	27·3	234
,, ,, ,, with 10% colophony	39·1	231·8
,, ,, ,, ,, 20% ,,	54·1	221·8
,, ,, ,, ,, 10% guaiacum	32·1	221·8
,, ,, ,, ,, 20% ,,	39·6	202·2
,, ,, ,, ,, 10% gallipot	42·3	221·8
,, ,, ,, ,, 20% ,,	56·8	211·1

Beckurts and Bruche have recorded the following values for seven samples of the genuine resin—

	Sp. gravity.	Acid value.	Ester value.
(1)	1·143	15·0	110
(2)	1·147	13·0	121
(3)	1·150	18·0	111
(4)	1·151	27·0	109
(5)	1·149	11·0	118
(6)	1·149	20·0	113
(7)	1·149	14·0	126

CASTOREUM

This valuable body is one of the few animal substances used in perfumery. It is described here, because it contains from 40 to 70 per cent. of resin, although not of vegetable origin.

Castor or castoreum consists of the dried membranous follicles of the beaver, *Castor fiber*, situated between the anus and the genital organs of both sexes. There are two pairs attached to each animal, the lower ones being pear-shaped and larger than the other pair. They contain an oily, viscid, highly odorous substance secreted by glands. The follicles are removed after the death of the animals and dried either by smoke or in the sun. When quite fresh castor is a white liquid of creamy consistence. Two varieties are met with commercially, the Canadian and the Russian. The Canadian castor is the one which is found in this country, the Russian variety only rarely reaching London. Castor is sold in the form of solid unctuous masses, contained in sacs of 2 to 3 in. long, much flattened and wrinkled, of a deep brown colour. A good sample should be powerful in odour and have a bitter nauseous taste. As far as the London market is concerned, castor comes on offer only once a year. It is practically a monopoly of the Hudson Bay Company, and the collections of castor during the year are offered by auction annually, towards the end of the year. The sale for 1916 was held in December, and the total amount offered was 1,073 lb.

Castor is still used to a very small extent as a drug, but only by old-fashioned practitioners. It is almost entirely employed in perfumery, acting as a very fine fixative of other odours.

Castor varies considerably in composition, the chief constituent being a resin, which is present to the extent of from 40 to 70 per cent. The characteristic odour is due to a small quantity of an essential oil. A peculiar crystalline principle named castorin is also present. Canadian castor contains between 4 and 5 per cent. of this body.

Mingaud gives the following analysis of castor—

Ethereal extract	88·4% (mostly resin)
Alcoholic ,,	0·8%
Aqueous ,,	0·1%
Acetic ,,	0·6%
Residue	2·2%
Volatile matter	7·9%
Ash	0·75%

Opopanax

Opopanax is usually stated to be derived from two entirely different sources, one known as Burseraceous opopanax, from *Balsamodendron kafal*, and the other as Umbellifer opopanax, from *Chironium opopanax*. There appears, however, reason to believe that the usual text-book statements are incorrect, and that what has been described as Umbelliferous opopanax is, in fact, liquid galbanum oleoresin, a product from Northern Persia.

The principal use of opopanax is as a perfume. It appears that the name opopanax was first given to a perfume material under a misapprehension. The *true* opopanax, which is not the perfume resin known by the name to-day, has a penetrating and somewhat offensive odour, recalling that of crushed ivy leaves. It has been employed to a small extent in medicine since the time of Dioscorides, and was probably, according to E. M. Holmes, in use long before his time. Up to about 150 years ago it was commonly employed in medicine, but its use for this purpose is now extinct. This substance is probably the product of the plant *Opopanax chironium*, a native of Greece.

At the time when opopanax perfume first became popular, a demand arose for the gum resin bearing that name, and the true, somewhat foul-smelling opopanax was imported. This odour led the early writers to class opopanax as a fetid resin, and Burton, in his *Anatomy of Melancholy* speaks of " opopanax, sagapenum, assafoetida, or some such filthy gums."

E. M. Holmes has done much to elucidate the question of the origin of the resin used in perfumery under the name opopanax, and draws attention to the strong similarity in appearance between the fetid resin, and the perfume resin, hence the ease with which confusion has arisen. The odour of opopanax perfume recalls that of bdellium, but with a more sweet, heavy fragrance. It is very similar to the perfumed bdellium or bisabol myrrh. Its exact botanical source was identified in the following manner. Some living pieces of the " myrrh " plants of South Arabia were sent home to Kew Gardens, where it was cultivated, and Mr. Holmes, whilst in one of the greenhouses there, noticed a drop of the gum resin exuded on the stem. He found this not to have the flavour of myrrh at all, but to have the characteristic flavour and odour of the bisabol or perfumed bdellium, and in his opinion this plant is the source of commercial perfume opopanax resin. The plant is clearly one of the *Burseraceae*.

Opopanax perfume resin is very slightly soluble in carbon bisulphide, insoluble in petroleum ether, and only a small proportion of gum is soluble in water. The confusion which exists as to the source of this product, obviously renders published analytical figures of very doubtful value. Dieterich gives the following figures, but it appears almost certain that samples Nos. 1 and 2 are not true perfume opopanax at all, whilst No. 3 may possibly be the resin in question—

	(1)	(2)	(2)
Acid value	32·4	35·0	53·4
Ester value	105·5	114·1	142·6
Saponification value	137·9	149·1	196·0

FRANKINCENSE

The gum resin, olibanum, is that which is known as frankincense. It was known to the Greeks under the name Libanos, and to the Romans as Olibanum, and to the Arabs as Luban, all of which are derived from the Hebrew word *Sebonah*, meaning milk. This gum resin has, from the earliest times been regarded as one of the indispensable ingredients of incense for religious purposes. E. M. Holmes quotes the following interesting account by Cosmas, as to the method adopted for trading between natives, and traders who were ignorant of their language. It relates more especially to the natives of the highlands of Abyssinia who collected the valued frankincense. " The gold caravan is usually made up of about 500 traders. With them they take a good quantity of salt and iron, and when they are close to the gold land, they rest awhile and make a great thorn hedge. Then they kill the cattle, cut them up, and split their joints upon the thorns, while they put out the salt and iron at the foot of the hedge. That done they retire to a certain distance. Now up come the natives with their gold in little lumps, and each places what he thinks sufficient above the beef, the salt, or iron which he fancies. Then they, too, go away. Next return the merchants and inspect the price offered for their goods. If content they take away the gold and leave the flesh, salt or iron thus paid for. If not content they leave the gold and other things together and retire again. A second visit is then paid by the blacks, and either more gold is added, or it is removed altogether, according as the purchaser thinks worth while."— (*Perfumery and Essential Oil Record*, 1916, 79.)

The trading in early days in frankincense appears to have been carried out on this basis.

There appear to be two varieties of frankincense known to the Arab collectors, one called Loban Dakar (or male frankincense) and Loban Maidi (or female frankincense). The trees yielding these two gum resins are *Boswellia carteri* and *Boswellia frereana* respectively. The former is found on the maritime limestone mountains, south of Berbera in Western Somaliland, and thence eastwards, whilst the latter is not found until the Habr Toljaela country is reached, and extends further west, into the country of the Warsangeli and Mijertain tribes.

The male frankincense is probably the only one found in European commerce, the female frankincense being used in the country as a kind of chewing gum. Three principal qualities are recognised on the London market (1) Fine tears, (2) ordinary and small lumps and tears mixed, (3) siftings.

Olibanum gum resin consists of a small amount of fragrant essential oil, together with gum which resembles gum arabic in chemical composition, and boswellic acid, both free and in the form of esters, with several complex bodies of unknown constitution.

It is rarely adulterated, although colophony, mastic, and sandarach resin are said to be used occasionally as adulterants. According to Kremel, three samples on analysis gave the following results—

		(1)	(2)	(3)
Resin		64%	72·1%	67%
Acid value	⎫	59·3	46·8	50·3
Ester value	⎬ of resin	6·6	41·0	60·5
Saponification value	⎭	65·9	87·8	110·8

This gum resin is employed almost exclusively for the purpose of incense manufacture.

LADANUM RESIN

Ladanum resin is a product of the greatest importance in the perfume industry, and has, of recent years become quite indispensable in the manufacture of particular types of perfumes possessing what is known as the " oriental " odour.

The principal plant yielding the resin is *Cistus ladaniferus*, but *Cistus cypricus* and *Cistus creticus* also yield ladanum resin. A good deal of high grade ladanum is produced in Spain, where no less than sixteen species of cistus have been recognised. The plant yielding the resin there, is *Cistus ladaniferus*, var. maculatus. The substance produced in Spain is pure ladanum, whilst much of the commercial resin was, and still is to a considerable extent, grossly adulterated. The Cretan product is very frequently almost entirely factitious.

According to Dieterich (*Analysis of Resins*, p. 196), at one time a quaint method was employed for collecting this resin, by driving flocks of sheep through the cistus shrubs and gathering the resin that adhered to the wool.

Ladanum occurs in dark-brown or black viscid masses, which are easily softened by handling. It exhibits a greyish fracture when broken, but the broken surface rapidly becomes black. It is not soluble in water, but dissolves almost completely in alcohol. The odour rather recalls that of ambergris. The adulterants are mixtures of other resins, principally common colophony. Very few samples of absolutely authentic origin have been examined, so that analytical figures must be accepted with some reserve, and the product must, to a very great extent, be judged by its odour value. Qualitative tests for such adulterants as colophony (*q.v.*) may be applied. The following analysis are by Dieterich (*loc. cit.*).—

	Acid value.	Ester value.
French Commercial	90·37	116·1
	91·98	120·3
,, ,,	98·1	102·1
	98·4	109·9
German commercial	54·1	167·9
	54·7	162·0
,, ,,	54·0	166·9
	51·9	168·4
Cretan commercial	113·8	87·9
	114·8	88·0

It is possible that all the above samples were more or less artificial products. This resin requires further investigation very urgently.

Cretan ladanum has recently been critically examined by E. J. Emmanuel, who found it to have the following composition—

Resin extracted by ether	48%
Resin extracted by alcohol after ether extraction	17%
Essential oil	2%
Ladaniol	0.8%
Reserves	15%
Gum	3.5%
Mineral matter	12%

Ladaniol was obtained from the ether extract by distillation after all the essential oil had been driven over. It forms white, fragrant crystals melting at 89° and may possibly be identical with champacol, the odorous substance obtained by Merck from champaca wood. According to Masson, the essential oil distilled from this valuable gum resin is a liquid of specific gravity about 0·950, boiling at 185° at 15 millimetres pressure. It contains acetophenone and other compounds not yet identified.

SAGAPENUM RESIN

This gum resin has, from time to time, been employed as a drug, but to-day is not met with in commerce very freely. It is the product of an umbelliferous plant, and comes from Persia. Apart from this, the source of the resin cannot be said to be known accurately.

The resin contains umbelliferone, and complex resinous substances of more or less unknown constitution. It occurs in the form of dark-brown masses, with white portions, rather brittle, but easily softening when handled. Its smell is not pleasant, recalling that of asafoetida. When shaken with hydrochloric acid, its solution in ether turns red-violet, and the resin gives similar reactions to galbanum, due to the presence of umbelliferone in both products. The acid and ester values of sagapenum resin are as follows—

Acid value.	Ester value.
12·0 to 17·0	29 to 42

TACAMAHAC RESIN

This resin is used to a limited extent in the manufacture of certain types of varnish, and is sometimes known as West Indian anime resin. It is the product of a number of trees, including *Icica heptaphyllum* and *Elaphyrum tomentosum*, which are mainly responsible for American and West Indian tacamahac resin, and *Calophyllum inophyllum*, which produces the East Indian variety.

Tacamahac resin is very similar to elemi. It is a yellowish-brown semi-transparent resin, with a slightly spicy odour. The particular variety which comes from Bourbon is rather soft and sticky, and is similar in character to the type of elemi known as caranna resin. A good deal of the tacamahac resin found in commerce is of very dubious origin and is probably a mixture of the genuine with other less valuable resins. The usual adulterants are cheap grades of elemi and similar resins and colophony.

On treatment with petroleum ether, from 45 to 60 per cent. is dissolved from the best grades. The iodine value of the resin varies from 68 to 80, so that adulteration with colophony is indicated by a high iodine value, as well as by a high acid value. Pure samples have an acid value which rarely falls outside the limits 25 to 40, and an ester value of 60 to 75, rarely as low as 40, except in specimens of very doubtful origin.

CHAPTER IV

TRUE GUMS

Gum Arabic

THE soluble gum produced in the Sudan has been an article of commerce ever since the first century of the Christian era. It was shipped to Arabian ports and thence to Europe, hence the name gum arabic. The name embraces gums from several botanical sources, but as most of the best gum of commerce is derived from species of Acacia, the name "gum acacia" is usually used to indicate the best varieties. In the Middle Ages, the trade was largely carried on through Turkish ports, and the gum was frequently known as Turkey gum. To-day, the names Sudan gum or Kordofan gum are commonly used.

The following account of the collection of the gum, and much of the information regarding the product, is due to the Director of the Imperial Institute (Bulletin No. 63). In the Sudan the best gum is collected from the grey-backed acacia tree, *Acacia senegal*, known locally as hashab. Inferior varieties are obtained from the red and white backed acacias, both of which are varieties of *Acacia senegal*, and are known locally as "talh" or "talha." A certain amount of gum is collected in the Blue Nile district, and there is a fair gum trade at Gedarif, which lies between the Blue Nile and Abyssinia; but the quantity of "hashab" gum produced there is hardly equal to that of the province of Kordofan, which is the principal seat of the gum-collecting industry. Kordofan lies to the west of the White Nile, almost 200 miles south-west of Khartoum. In this province the gum is transported either direct to Khartoum by camels or to Goz Abu Guma and El Dulime, towns on the White Nile, and is there put into boats. The greater part of the gum was formerly dried and cleaned at Omdurman, which lies on the Nile opposite Khartoum. But at present only about 8 per cent. of the production is treated there, and the rest is sent direct to Cairo and Suez or Port Said, whence it is exported to Europe, etc. At least one-half goes to Suez.

The gum loses about 15 per cent. by evaporation between the region where it is gathered and the port of export. There is also an export of gum from Suakim, and it is probable that with the completion of the Berber-Port Sudan railway, a much larger proportion of the gum exports will pass *viâ* this route to the Red Sea and the Mediterranean.

In Kordofan the gum is obtained both from gardens of acacia trees, which are private property, and from wild or unowned trees; the first kind is known as " hashab geneina " (*i.e.*, garden hashab), and the second, which is of less value, is known as " hashab wady." The latter exudes naturally from the trees, and is slightly darker in colour; it is usually in pear-shaped pieces of variable size proportionate to the length of time between successive collections. A dirty gum which is sometimes found exuding is known as " kadab," and is rejected.

The conditions favourable to the production of gum are a ferruginous sandy soil, with a good natural drainage, and probably a moderately heavy rainfall during the rainy season is beneficial, and dry heat during the collecting season. Excessive moisture in soil otherwise suitable appears to prevent the production of gum.

In the " geneinas " gum is obtained by artificially

incising the trees. Soon after the end of the rains, bark is removed in strips from the principal branches of all trees in the garden of 3 years old and upwards; the strips should be 1 to 3 in. wide, according to the size of the branch, and 2 to 3 ft. in length. They are removed by cutting the bark with an axe and then tearing off by hand. The incision should not penetrate into the wood, and a thin layer of the liber or inner bark should be left covering the wood. About sixty days afterwards, the first collection of gum is made, and after that the garden is completely picked over every fourth day until the rains recommence and new leaves appear on the trees; at this stage the exudation ceases. In Kordofan the rainy season ceases at the end of September, and recommences in the middle of June. Young hashab trees, 8 to 10 ft. high and 6 to 8 in. in girth, will produce gum, and the limits of age may be taken as 3 to 15 or 20 years; probably trees of 8 to 12 years are the most productive.

" Talh " or " talha " gum is chiefly collected in the forests of the Blue Nile. There are two varieties of the talha acacia tree, *Acacia seyal*; the bark of one is covered with a red powder and that of the other with a white powder, and they are consequently known as " red " and " white " talha respectively. Both varieties produce gum, but the red talha is more abundant than the white, and consequently most of the talha gum is derived from that variety. The talh trees are said not to be barked or wounded by the collectors who gather the gum they find exuding. The gum is cleaned from pieces of bark and other debris at Omdurman or Khartoum, and a small proportion of it is picked and dried by exposure to the sun on the banks of the Nile, and exported as " picked gum." Most of the gum is, however, exported in the mixed condition and is sorted at European centres, of which the most important has hitherto been Trieste.

The method of collecting this gum is as follows. The gum exudes from the stem and branches spontaneously, and the flow is usually stimulated by making incisions in the bark. The method of tapping consists of cutting the bark with a small axe, and tearing off a thin strip about 2 or 3 ft. in length and 1 to 3 in. in width according to the size of the branch. The exuded gum hardens on exposure to the air and is then collected, dried and exported. The gum is official in the British Pharmacopoeia which requires it to have the following characters—

In rounded or ovoid tears or masses of various sizes, or in more or less angular fragments with glistening surfaces; nearly colourless, or with a yellowish tint. Tears opaque from numerous minute fissures; very brittle, the fractured surface being vitreous in appearance. Nearly inodorous; taste bland and mucilaginous. Insoluble in alcohol (90 per cent.); almost entirely soluble in water, the solution being translucent, viscous and slightly acid. When dissolved in an equal weight of water, the solution is not glairy, and after admixture with more water, yields no gummy deposit on standing. An aqueous solution (1 in 1) exhibits slight laevo-rotation (absence of dextrin, certain sugars, etc.), 10 millilitres of the same solution are not precipitated by solution of lead acetate; are not, after previous boiling and cooling, coloured blue or brown by 0·1 millilitre of $N/10$ solution of iodine (absence of starch and dextrin) or bluish-black by T. Sol. of ferric chloride (absence of tannin). Ash not more than 4 per cent.

Gum arabic is composed essentially of the calcium salt of arabin or arabic acid, which is obtainable in a pure state by dialysing a solution of the gum previously acidulated with hydrochloric acid. The glue-like liquid thus obtained is laevo-rotatory and is not precipitated by pure alcohol, but is thrown down if traces of salt or acid are present. After evaporation to dryness and heated to 100° the arabin does not dissolve again, even in hot water, but swells up into a gelatinous mass, which dissolves gradually when treated with soda, lime or baryta water, and yields a liquid which is indistinguishable from the aqueous solution of ordinary gum arabic.

Most varieties of gum arabic—which include the Sennaar, Senegal, East Indian and Levantine—are laevo-rotatory, whereas Australian gum is frequently optically inactive, while Gedda gum is dextre-rotatory. Chemically, these gums are analogous to the laevo-rotatory varieties.

The inferior qualities of gum contain a small amount of a reducing sugar, which is removable by treating with alcohol.

The specific gravity of air-dried gum arabic varies from 1·35 to 1·49, but when it is completely dried at 100°, loses about 13 per cent. of water, the density increasing considerably.

Gum arabic is nearly odourless and has a mucillaginous and insipid taste. It dissolves slowly in about twice its weight of water, forming a thick transparent mucilage of acid reaction. The gum is somewhat soluble in dilute spirit but is quite insoluble in any liquid containing more than 60 per cent. of alcohol, and is precipitated from its aqueous solution if a large proportion of spirit is added.

The aqueous solution of gum arabic is not precipitated by neutral lead acetate, but with basic acetate forms a white jelly. Its solution is also precipitated by potassium or sodium silicate, borax, ammonium oxalate, mercuric chloride and ferric salts.

The following figures illustrate the importance of the

Sudan gum trade, and the share in the exports of Sudan gum taken by the United Kingdom—

EXPORTS OF SUDAN GUM FROM EGYPT.

Year.	Kilos.	Value £ (E).
1885	1,146,879	97,671
1890	7,052	469
1895	149,955	5,856
1900	1,863,072	93,847
1905	8,838,483	217,132
1906	7,689,834	157,330

According to the reports of the Secretary to the Sudan Economic Board for 1907 and 1908, the total exports of the gum from the Sudan for these two years were valued at £E154,592 and £E175,269 respectively.

IMPORTS OF GUM FROM EGYPT TO UNITED KINGDOM.

Year.	Cwt.	Value £.
1903	43,334	82,370
1904	32,879	47,168
1905	27,881	41,995
1906	25,599	35,333
1907	38,579	62,530

These figures indicate the necessity for careful examination of the product. According to A. H. Allen (*Allen's Commercial Organic Analysis*) the following scheme is of importance for the examination of gum arabic.

Gum arabic should not contain more than about 4 per cent. of ash. It should be soluble almost without residue in cold water. The solution should be free from starch and dextrin, as indicated by the negative reaction with iodine solution; but should be rendered turbid with oxalic acid, which the solution of dextrin is not. The better varieties of gum arabic do not reduce Fehling's solution when heated to boiling with it, any red precipitate being due to the presence of a reducing sugar, small quantities of which exist naturally in certain inferior kinds of gum, though any considerable amount would probably have been introduced as an impurity in an admixture of dextrin.

According to Z. Roussin (*Jour. de Pharmacie*, (4) vii. 251), gum arabic and dextrin may be distinguished and separated by means of ferric chloride, which precipitates only the former, and the resultant precipitate washed with rectified spirit and dried. 1 grm. of the dry residue is then dissolved in 10 cc. of water, the solution mixed with 30 cc. of proof spirit, 4 drops of ferric chloride solution (containing 26 per cent. of the anhydrous chloride) added, followed by a few decigrammes of powdered chalk; and after stirring briskly and leaving the liquid at rest for a few minutes it is then filtered. The precipitate is washed with proof spirit, and the dextrin is precipitated from the filtrate by the addition of very strong alcohol. After twenty-four hours the spirituous liquid is decanted, the dextrin dissolved in a small quantity of water, the resultant solution evaporated at 100°, and the residue weighted. The precipitate containing the gum must be dissolved in dilute hydrochloric acid, the arabin precipitated by adding absolute or very strong alcohol, and after being washed with spirit is dissolved in water, the solution evaporated, and the residue weighed. The precipitation of gum arabic from a dilute alcoholic liquid by ferric chloride and chalk is so complete that nothing but calcium chloride can be found in the filtrate, while the precipitate similarly produced in a solution of dextrin is perfectly free from the latter body. By the formation of a cloud on adding ferric chloride alone, the presence of gum arabic is sufficiently demonstrated, while the clouding of the filtrate from the iron-chalk precipitate on addition of alcohol proves the presence of dextrin.

A large proportion of dextrin would be indicated by he dextro-rotatory action of the solution, but the variation in the optical activity of both natural gum arabic and commercial dextrin would prevent the quantitative application of the test.

To separate gum arabic from sugar, Andouard dilutes 10 grm. of the syrup with 100 cc. of alcohol of ·800 specific gravity, adding twenty drops of acetic acid and stirring vigorously. After three hours the liquid is poured on a double filter, when the gum forms a cake which readily drains. This is dissolved in a little water, and the precipitation repeated, the precipitate washed with alcohol, dried at 100° and weighed. It is then exposed to the atmosphere for twenty-four hours, when it will have taken up its normal amount of moisture. The inferior varieties of gum are employed on a large scale as thickening agents in calico-printing. Good gum neither tarnishes nor alters delicate colours and does not weaken the mordants. The action of gum on delicate colours may be ascertained by printing a solution of the sample mixed with cochineal-pink or fuchsine upon pure wool. The material is then steamed and washed, when, if the gum is pure, no trace of yellowness will be apparent. Too great an acidity of the gum gives it a solvent action on mordants, and hence renders it unsuitable for use.

The relative viscosity of samples of gum is an important character in judging of their quality. This may be tested by making solutions of 10 grm. of each sample in a little warm water, diluting the liquids to 100 cc. and ascertaining the rate at which the solutions flow from a glass tube drawn out to a fine orifice. A recently prepared solution of gum of the best quality should be used as a standard.

The following analyses are those on a number of samples carried out in the laboratories of the Imperial Institute.

	Sudanese gum, Hashab, 1914.	Senegal gum, "Gomme du bas du fleuve."	Sudanese gum, Specially selected Hashab, 1903.	Senegal gum, "Gomme petite blanche."	Senegal gum, "Gomme grosse blonde."	Sudanese gum, Gezira, 1903.	Sudanese gum, Tâlh, 1903.
Moisture, per cent.	13.2	16.10	11.3	16.1	16.0	12.4	12.2
Ash, per cent.	3.1	3.5	3.3	3.0	3.1	2.7	2.6
Dry matter soluble in water, per cent.	86.5	82.0	87.6	80.6	83.0	87.2	85.2
Acidity (Milligrams of potash required per gramme of gum)	2.4	1.9	1.2	0.8	1.2	2.0	2.8
Viscosity of 10 per cent. Solution	31.4	22.5	16.3	32.4	28.7	18.7	25.0
Character of mucilage	Clear, very pale brown. No marked taste or odour.	Opaque, dark brown colour. Slightly bitter.	Clear and almost colourless. No marked taste or odour.	Clear, faintly yellow. No marked taste or odour.	Clear, slightly yellow. No marked taste or odour.	Clear, pale yellow. Slight sour odour.	Clear, pale reddish-brown. Slight burnt taste.

These results indicate that the Senegal gums contain more moisture than the Sudan gums. The greater brittleness of the Sudan gums is due to their drying and becoming permeated by a large number of fissures. The conclusion reached by the Imperial Institute on these gums was as follows.

"The most important difference between the two classes of gums are, however, shown by the colours and the viscosities of their mucilages. On comparing the 'Hashab gum of 1904' and the 'Gomme du bas du fleuve,' which are both natural unpicked gums, it will be seen that the former is much lighter in colour than the latter, a feature which is to the advantage of the Sudan gum, since absence of a marked colour is a necessity for a number of manufacturing purposes to which gums are applied. On the other hand, the viscosities, that is roughly, the 'strengths' of the Senegal gums are, on the whole, higher than those of the Sudan products. This difference is very noticeable when the specially selected 'Hashab of 1903' is compared with the selected 'Gomme petite blanche.'"

"In reporting the results of this comparison of Senegal and Sudan gums to the Government of the Sudan, it was pointed out that though it was unsafe to draw general deductions from the comparison of such a small number of samples, yet there appeared to be some ground for the opinion that Senegal gum was for some purposes superior to the Sudan product, though the latter had the compensating advantages of being cleaner and of lighter colour. A number of suggestions were also made as to the necessity of systematically examining the gum produced from year to year in the Sudan, so that data could be accumulated for the solution of questions of this kind as they arose, and the suggestion was made that it might be desirable to classify Sudan gum into a larger number of grades before export than at present."

The Sudan gums were submitted for trial to a firm of manufacturing confectioners, who described the "Hashab gum of 1903" (specially selected and dried) as a white, clean gum, yielding a very pale, clean, viscous solution and of good flavour; and the "Hashab gum of 1904" as consisting of fine, bold nodules, free from dirt and giving a pale, highly viscous solution, of good flavour and odour, and therefore of special value to confectioners. The "Gezira gum of 1903" was described as yielding "a somewhat darker but still satisfactory solution, fairly viscous, with a sourish smell but good flavour."

A similar gum has been collected in the Senaar forests and found to resemble very closely the ordinary Sudan gum. The colour is pale-yellow, but the odour and taste are slightly unpleasant. An examination of two samples gave the following results at the Imperial Institute laboratories—

Moisture .	14·2	12·2
Ash, per cent. .	3·91	2·66
Portion soluble in water, per cent.	84·6	85·2
Portion insoluble in water, per cent. .	1·2	2·6
Acid No. .	1·8	1·8
Reducing power	slight	very slight.

Morocco exports about one hundred tons of gum annually, which is probably derived from *Acacia arabica* and *Acacia gummifera*. It does not differ materially from Sudan gum arabic.

The gum industry of the French colony, Senegal, is of much more recent origin than that which has existed for centuries along the Nile valley. The gum is obtained principally from *Acacia Senegal*, although other species contribute to the output of Senegal gum,

The gum is collected by Moors during the months of December, January and February, and again in April onwards to July. The Moors barter it with French merchants who send it *viâ* St. Louis Rufisque and Freetown to Europe.

Three qualities of crude Senegal gum are produced. They are described as follows by the Director of the Imperial Institute—

1. *Gomme du bas du fleuve.*—This quality is produced in the district of Podor in Lower Senegal. It is the best of the Senegal gums, and occurs in large rounded or thick vermiform tears. Its colour varies from almost white or pale sherry tint to brownish-yellow.

2. *Gomme du haut du fleuve.*—This variety is obtained in Fouhlah-land, Guidimaka, and Bambouk, all in Upper Senegal. It ranks second in price, and occurs in rounded, vermiform or branched tears, smaller in size than the first quality, and on the whole darker in colour.

3. *Gomme friable, Salabreda,* or *Sadra beida.*—This, the poorest quality of Senegal gum, consists of small grains (showing a tendency to cohere into masses) and small vermiform tears. The latter are usually only slightly coloured, but the grains are brown.

Senegal gum is almost entirely exported to France.

The average value of the exports of Senegal gum is from £50,000 to £70,000 per annum, of which almost the whole is sent to France, the quantity reaching this country rarely, if ever, exceeding about £1,000 in value.

A fair amount of gum, also principally from *Acacia Senegal*, now reaches this market from Northern Nigeria, the average annual value being about £8,000, and possibilities exist in this direction in the Gold Coast Colony, Orange River Colony, and various other parts of Africa.

GHATTI GUM

Ghatti or ghati gum is the name applied locally to a gum produced in India, but, as a good deal of gum is imported from neighbouring sources into India, mixtures of gums from various sources are understood by the term in British commerce. Ghatti gum is much less soluble than gum arabic, but also yields a highly viscous mucilage. An exhaustive examination of this gum, in specimens of known origin, has been made by the Imperial Institute chemists, who have published the series of analyses of nine samples of unmixed gum shown on the next page.

Rideal and Youle (*Year Book of Pharmacy*) have made an exhaustive examination of these Indian gums in comparison with ordinary samples of gum arabic. They found that, whilst a good gum arabic yielded the mucilage official in the British Pharmacopoeia by using 1 part of gum with 2·5 parts of water, a mucilage of ghatti of the same viscosity was yielded by using 1 part of ghatti gum with 8 parts of water. The ghatti mucilage, however, must be strained from the insoluble matter present, which appears to consist of metarabin. The following reactions are given for the two gums, from which it is apparent that alcohol, ammonium oxalate, ferric chloride and mercuric chloride are useful reagents for differentiating between the two classes of gum.

	Reagent.	With Ghatti.	With Gum Arabic.
1	Ammonium oxalate.	Slight turbidity.	Copious white ppt.
2	Lead basic acetate.	Slight precipitate.	Copious gelatinous ppt.
3	Ferric chloride.	Slight darkening, gelatinous ppt.	No darkening, no gelatinous ppt.
4	Borax.	Gelatinizes.	Does not gelatinize.
5	Stannous chloride.	Bleaches, no gelatinizing.	Bleaches.
6	Alcohol (equal bulk).	Slight precipitate.	Copious precipitate.
7	Mercuric Chloride.	White stringy ppt.	No reaction.

RESULTS OF EXAMINATION.

	Acacia Jacquemontii.			Prunus eburnea.	Elaeodendron glaucum.	Acacia catechu.		Acacia arabica.	Acacia Farnesiana.	Acacia modesta.	Acacia Senegal.		Anogeissus latifolia.
	From Amritsar.	From the Punjaub.	From the Punjaub.	From Baluchistan.									
	Per cent.	Per cent.	Per cent.	Per cent.	Per cent.	Per cent.	Per cent.	Per cent.	Per cent.	Per cent.	Per cent.	Per cent.	Per cent.
Moisture	18·5	18·5	14·8	14·8	15·2	15·7	16·8	15·0	13·8	16·2	16·3	15·5	16·1
Ash	3·53	3·5	2·84	2·54	2·59	3·24	2·87	2·27	1·83	2·69	2·68	2·41	3·11
Dry matter soluble in water.	80·02	81·9	84·0	80·7	82·0	82·4	81·2	78·6	85·40	85·0	84·0	78·2	71·5
Acidity No.	4·0	3·2	4·8	0·8	—	—	—	—	—	—	—	—	—
Character of Mucilage	Faintly opaque, slightly yellow, adhesive.	Clear, faintly yellow.	Slightly yellow.	Fine clear, colourless adhesive solution.	Clear, very slightly yellow.	Clear, light reddish-brown.	Light yellowish-brown.	Clear, pale reddish-brown.	Clear, pale yellowish-brown.	Clear, pale yellow.	Clear, faintly yellowish-red.	Viscid, pale yellowish-brown.	Viscid yellow solution.

Useful information was yielded by treatment with alcohol. Five grammes of the gum were dissolved in 20 cc. of water, and the solution filtered from insoluble residue. To the cold solution 90 cc. of 95 per cent. alcohol was added, and the precipitate washed with 30 cc. of alcohol of the same strength. It was then dried and weighed, then redissolved in water and its rotatory power determined. With two samples of gum arabic and two of gum ghatti, the following results were obtained—

Sample.	Weight gum taken.	Weight alcohol precipitate.	Weight gum in filtrate.	$(a)j$ Original gum.	$(a)j$ Alcohol ppte.	$(a)j$ Filtrate.
Arabic 1	grms. 5·000	3·1872	1·4162	+66·2°	+57·9°	+55·1°
,, 2	5·000	2·3312	2·4108	−38·2°	−20·1	−64·9°
Ghatti 1	3·245	·4065	2·8385	−140·8°	−106·0°	72·4°
,, 2	2·255	·2900	1·8078	+147·06°	−106·04°	−69·0 (*loc. cit.*)

The ghattis are generally laevo-rotatory, and the alcohol precipitate is apparently of a different kind to that yielded by gum acacias; that is, the precipitate is more laevo-rotatory than the filtrate, while the opposite is the case with gum arabics, whether they are laevo-or dextro-rotatory. Both classes of gums, however, it would appear from these experiments, consist of at least two kinds of gum, one of which is more soluble in alcohol than the other and differs in its action on polarized light. Similar work done by O'Sullivan with pure arabin points to the same conclusion.

These results are interesting from a pharmaceutical point of view, as it may be found possible to obtain from ghatti gum, by fractional precipitation with alcohol, a gum which will be identical with ordinary gum arabic. Rideal and Youle are still working in this direction.

Australian Gums

A considerable amount of gum "arabic" is collected in the Australian Commonwealth, from species of *Acacia* known as wattles—the gum being usually known as wattle gum. Most of it is very pale in colour and suitable for the manufacture of adhesive mucilage. It does not present any particular differences from ordinary acacia gum or gum arabic that call for notice.

There are, however, a number of other Australian gums, which have been investigated, principally by the Imperial Institute, and which, although not commercial articles of any importance, show sufficient possibilities to be noticed here. The only ones to which attention will be drawn are the following—

1. Gum from *Macrozamia perowskiana*.
2. ,, *Macrozamia spiralis*.
3. ,, *Ceropetalum gummiferum*.
4. ,, *Ceropetalum apetalum*.
5. ,, *Flindersia maculosa*.

The gum from *Macrozamia perowskiana* is found in New South Wales in flattened pieces somewhat resembling ordinary button lac, but much paler in colour. It absorbs water in the same way as does gum tragacanth, swelling to about one hundred times its original size. It then forms a perfectly transparent jelly. The gum from *Macrozamia spiralis* is found in New South Wales and Queensland, and is soft when collected, but soon hardens to scaly pieces, which behave towards water in the same manner as the gum from *Macrozamia perowskiana*. The two gums have the following compositions—

Arabin	·94	1·07
Metarabin	77·22	71·7
Sugar	1·02	1·1
Water	14·81	21·71
Ash	6·66	4·72

These gums resemble cherry gum, and to a small extent gum tragacanth. The gums of *Ceropetalum gummiferum*, the "Christmas Bush" of New South Wales is exuded from the cut ends of the wood and forms tears of a fine ruby-red colour, or cakes which have little colour, but which impart a rich orange-brown colour to water. The gum of *Ceropetalum apetalum* is similar, but has a marked odour of coumarin, which is present in considerable quantities in the bark of the tree.

Both these gums contain tannic acid, but also considerable quantities of actual gum. They are therefore intermediate in character between the true gums and the kinos, which are very often referred to as "gum kinos," although they are essentially tannin compounds, and are therefore not described here.

Maiden considers *Ceropetalum apetalum* as worthy of note as an available source of coumarin, and states that the presence of that substance sharply separates the two gums. The following difference also appears to be constant. The ash of *Ceropetalum gummiferum* is quite white, while that of *Ceropetalum apetalum* is dark brown, very bulky, and difficult to ignite. It contains but a small percentage of iron, but manganese is abundant. The composition of the gums is shown in the following table—

	C. gummiferum.	C. apetalum.
Tannic acid, estimated as gallotannic acid	16.76	6.35
Phlobaphenes (soluble in alcohol)	19.5	12.21
Phlobaphenes (insoluble in alcohol, together with metarabin)	41.6	52.09
Coumarin	nil (variable)	2 to 3
Accidental impurity	2.5	2.0
Moisture	16.7	20.47
Ash	1.8	3.44

The gum from *Flindersia maculosa*, the so-called

leopard tree of the interior of New South Wales and Queensland, is obtained as an exudation from the stems and branches during the summer months. It makes an excellent adhesive mucilage, and is also used by the aboriginals as food. Maiden has examined the gum and gives the following analyses of two samples, from which he draws the conclusion that the leopard tree gum is, to all intents and purposes, a good quality gum arabic. His figures are as follows—

	(1)	(2)
Arabin	80·2%	80·1%
Metarabin	nil	nil
Water	16·49%	16·4%
Ash	2·76%	2·63%

Persian Insoluble Gum

There are several gums which are intermediate in character between the gums of the arabic type and gum tragacanth, in that they are what is colloquially known as "semi-insoluble," that is, they form thin jellies, instead of true solutions or stiff semi-solid jellies. The typical gum of this class is the so-called Persian insoluble gum, which is exported to a considerable extent from Basra and other ports in the Persian Gulf. Very little is known of the botanical source of these gums, but the principal tree which provides the exudation is probably *Amygdalus lesocarpus*. The trade is considerable, the exports from Bushire in some years reaching the value of £65,000.

A similar gum is found in Northern Nigeria and has been examined by the Imperial Institute. It was found to have the following characters—

Moisture	15·4%
Ash	2·42%
Amount soluble in water	76·6 %
Acidity	0·0

The mucilage obtained by solution of the gum in water was precipitated by alcohol and by a solution of basic lead acetate, but not by solution of ferric chloride, and in these respects it resembles mucilage prepared from gum arabic, but, unlike the latter, it was only slightly adhesive when applied to paper. The insoluble portion of the gum swelled into a translucent jelly in contact with water.

A similar product from the Gold Coast Colony has also been examined and found to have the following composition—

Moisture	13·6%
Ash	4·6%
Acidity	10·6%

It forms a rather sour and brownish coloured mucilage.

To conclude the section of the gums proper, the following notes on the chemical constituents of a number of the typical soluble gums, by Meininger (*Arch. Pharm.* 1910, 248, 871), which will be of interest to those desiring more chemical knowledge of the gums than can be gone into in detail in so limited a work as the present. The author acknowledges the most valuable information in regard to the above series of gums to the publications of the Imperial Institute.

Gum of Acacia pycnantha.—Moisture, 13·55 per cent.; ash, 0·92 per cent., of which 0·28 per cent. was Ca and 0·123 per cent. Mg.; insoluble matter, 0·64 per cent.; $a_D - 19.39°$. The arabinic acid isolated from this gum contained 43·44 per cent. of C, 6·24 per cent. of H, 50·32 per cent. of O, and 1·31 per cent. of N. The N content of the original gum is 2·19 per cent. On hydrolysis 58·61 per cent. of galactone, 16·98 per cent. of pentosane, and 2·92 per cent. of methyl pentosane, are obtained. The greater part of the gum is an arabogalactane. *Gum of Acacia horrida.*—Moisture, 15·34;

ash, 2·59 per cent., including Ca 1·06 and Mg. 0·345 per cent.; insoluble matter, 0·98 per cent.; a_D +53·94. The arabinic acid gave: C, 44·67; H, 6·19; O, 49·14; N, 0·71 per cent. The N-content of the original gum was 1·51 per cent. Hydrolysis gave pentosane 36·5; methylpentosane 2·82, and galactane 27·36 per cent. *Gum of Acacia arabica.*—Moisture, 14·39; ash, 2·41, containing Ca 0·765 and Mg. 0·106 per cent.; N. 1·39 per cent. Hydrolysis gave pentosane 50·43 and galactane 21·85 per cent. *Gum of Melia azadirachta.*—Moisture, 15·41; ash, 2·99, containing Ca 0·76 and Mg. 0·294 per cent. Insoluble matter, 0·27 per cent.; a_D − 57·16°. Hydrolysis gave pentosane 26·27; galactane, 11·11 per cent. The galacto arabane of the gum consists of laevoarabinose and dextro-galactose in the proportion of 1 : 2. The gum contains 4·49 per cent. of N. In addition to the above gums, the percentages of N found in the following were: *Acacia adans Onii*, 1·93; *A. senegal*, 1·81; *Feronia elephantum*, 1·53; *Anacarelium occidentale*, 0·92 per cent.

GUM TRAGACANTH

Tragacanth is a gum obtained by the exudation from the stem of *Astragalus gummifer*, and other species of *Astragalus*, small shrubs widely distributed throughout the Turkish Empire and Persia. It appears to be produced by the process known as gummosis of the cell walls in the pith and medullary rays. It swells by absorbing water and on account of the pressure in the interior of the stem finally forces itself out through cracks or through artificial incisions which are made to increase the flow. It is collected when dry and graded for market. The finest gum which has been exuded from the long incisions dries almost white in colour, and in flakes and is known as " flake " Tragacanth, being graded according to appearance. The portions which are forced through more or less rounded holes, and which dry in tears or vermiform pieces, are known as "vermicelli" tragacanth. The more inferior qualities are known as "hog" tragacanth. Two varieties of "flake" tragacanth are found on the London market, viz., the Persian or Smyrna varieties. Persian tragacanth occurs in thin, horny, translucent flakes. The Smyrna variety is more opaque and less ribbon-like. "Hog" tragacanth appears to be little good and is the gum obtained from a species of *Prunus* and known as caramania gum. It occurs in yellowish or brown opaque pieces, and it resembles the genuine tragacanth in most respects.

The composition of tragacanth has not yet been fully investigated, but the part soluble in water appears to be a complex acid, which, on hydrolysis, yields various sugars and geddic acid, whilst the part insoluble in water consists of a complicated acid which breaks down, on hydrolysis, into sugar and bassoric acid. Traces of starch and cellulose are also found in the gum.

Tragacanth is employed in medicine chiefly as a suspending agent in mixtures containing volatile oils, resins, or heavy insoluble powders. It is official in the British Pharmacopoeia, which requires it to have the following characters: Thin flattened flakes, irregularly oblong, or more or less curved, marked on the surface by concentric ridges. Frequently $2\frac{1}{2}$ centimetres-long and 12 millimetres wide. White or pale yellowish white, somewhat translucent. Horny, fracture short. Inodorous; almost tasteless. Sparingly soluble in water, but swelling into a gelatinous mass, which may be tinged violet or blue by decinormal solution of iodine. Ash not more than 4 per cent.

According to Giraud, gum tragacanth contains, on

an average, about 60 per cent. of a pectinous compound, which yields pectic acid on boiling with water containing a trace of hydrochloric acid. It also contains, according to the same authority, 8 to 10 per cent. of a soluble gum of the nature of arabin, 5 to 6 per cent. of starch and cellulose and 3 per cent. of mineral matter. The average amount of moisture is 20 per cent.

Gum tragacanth is very hard to powder, and is best made into mucilage by soaking the pieces in fifty times its weight of water, when it swells up into a thick jelly-like mucilage without actually dissolving. When diffused in a much larger amount of water it forms a ropy liquid which can be filtered. Mucilage of tragacanth is coloured yellow by solution of caustic soda; a solution of the gum gives no appreciable precipitate with borax, alkaline silicates or ferric chloride, but is precipitated in clots by alcohol. Solution of lead acetate thickens it, and, on treating the mixture, throws down a precipitate of the gum acids combined with lead.

The cheaper varieties of gum tragacanth are used in the calico printing industry, for which purpose the gum is first soaked in water for twenty-four hours until it has swelled to the fullest possible extent. It is then boiled with more water for about six hours, when a thick homogeneous solution results, but which has not a great deal of cohesive power.

Gum tragacanth is sometimes adulterated with cheaper gums, when in the powdered condition. The commonest adulterant met with is powdered gum acacia.

According to Reuter, if powdered tragacanth be extracted by means of 95 per cent. alcohol, and the liquid evaporated, the residue contains a little fat, a bitter principle and a trace of sugar.

The tragacanth of commerce is principally obtained

from the mountainous regions of Asia Minor, Syria, Armenia, Kurdistan and Persia.

The following are the principal species known to yield the gum—

1. *Astragalus gummifer*, a small shrub widely distributed in Syria, Armenia and Kurdistan.

2. *Astragalus adscendens*, a shrub growing to about 4 ft. in height, and found in South Western Persia at altitudes of 9,000 to 10,000 ft. It is also found in Armenia and Kurdistan.

3. *Astragalus leioclados*, found in Persia.

4. *Astragalus brachycalyx*, a shrub, 3 ft. in height, found on the mountains of Persian Kurdistan.

5. *Astragalus microcephalus*, a widely distributed shrub, found all over Asia Minor and Armenia.

6. *Astragalus pycnocladus*, a Persian shrub, said to yield aboundant supplies of the gum.

7. *Astragalus stromatades*, found chiefly in Asia Minor.

8. *Astragalus Kurdicus*, a native of Silicia and Cappadocia.

9. *Astragalus verus*, found in Persia and Asia Minor.

10. *Astragalus parnassi*, a small shrub found on the northern mountains of the Morea.

In July and August, the shrubs are stripped of their leaves, and short longitudinal incisions or slits are made in the trunks. According to a British Consular Report on the trade of Kermanshah, 1903–1904, No. 3189, page 28, " the top of the plant is burnt, and when the leaves are all consumed the fire is put out and incisions are made." The gum flows out, and, drying spontaneously, is ready for gathering in three or four days. If the weather is fine during the drying process, the " white leaf " form of gum is obtained; this is the most prized variety. If, on the other hand, rain falls, or the wind rises, particles of dust are carried into the surface of the gum which thereby loses its whiteness, and becomes the "yellow leaf" form, the second quality. The shape of the incision, of course, determines the form of the pieces; longitudinal incisions produce "leaf" or "flake" tragacanth, punctures yield "vermicelli" tragacanth, while irregularly-shaped incisions give knob-like masses, generally coloured, and of relatively low value. Another form, known in Persia as "Arrehbor," exudes from branches, which have been cut with a saw. In Persia the productive life of the shrub is seven years.

Smyrna is an important market for gum tragacanth; it is conveyed to that port of native dealers, who purchase it from the peasants, in bags containing about 2 quintals each. It is there sorted into the various qualities in order to fit it for the European market, packed into cases containing about 2 cwt. and shipped to London, Marseilles, or Trieste. Basra (near the mouth of the Euphrates) is also an important port of shipment. —(*Colonial Reports—Imperial Institute.*)

The exports of tragacanth from Smyrna are given in the following table—

EXPORTS OF GUM TRAGACANTH FROM SMYRNA.

Year.	Cwts.	Value in £.
1901	1,660	4,040
1902	3,000	9,577
1903	2,600	6,237
1904	2,300	5,104
1905	1,180	8,165
1906	880	5,369

There are a number of "insoluble gums" which closely resemble gum tragacanth in general characters, which have been examined in the laboratories of the Imperial Institute. A gum from Nyasaland, whose botanical origin has not been identified was found to have the following characters. It consists of small fragments of translucent gum, varying in colour from pale yellow to deep brown. It had a slight odour of acetic acid and was almost tasteless. On analysis it gave the following results—

Amount soluble in water	32·8%
Moisture	15·6%
Mineral matter	2·57%

The portion insoluble in water swelled up to a translucent jelly, similar to that of gum tragacanth.

The author does not agree with the statement made by the authorities of the Imperial Institute that insoluble gums of this class have at present no commercial value unless they can be obtained, like the well-known insoluble tragacanth gum, almost free from colour. Dark-coloured low grade tragacanth commands a market for certain purposes, where colour is of no importance at all, for example, in the manufacture of fumigating pastilles and other articles, where fine powders require " binding " together.

A sample collected in the Bukedi district of Uganda was examined. It is a gum locally known as " Nongo," and is derived from a small tree which has been identified as *Albizzia brownei*. On analysis it gave the following results—

Moisture	16·9%
Mineral matter	4·6%
Dirt	2·7%

In order to effect any appreciable solution in water, it was found necessary to allow a small amount of the powdered gum to remain in contact with a large volume of water for five or six days with continual shaking. The solution so formed was rather gelatinous, and was so viscid that a 1 per cent. solution was found to have approximately the same viscosity as a 20 per cent. solution of Sudan gum acacia. A 10 per cent. " solution " furnished a thin jelly in which a proportion of the gum was only swelled up without dissolving.

INDEX

ABIETIC acid, 15
Acacia arabica, gum of, 88
—— gummifera, gum of, 88
—— horrida, gum of, 96
—— pynantha, gum of, 96
—— Senegal, 78, 88
Acaroid resins, 19
Albizzia browner, 102
Alpha-amyrin, 9
Amber, 17
——, analytical values of, 18
Ammoniacum, 35
—— analytical values of, 36
Amygdalus lesocarpus, 95
Amyrin, 11
Amyris elemifera, 9
—— Plumieri, 9
Analysis of shellac, 28
Animi, 1, 2
Arrehbor, 101
Asafoetida, 32
Asaresinotannol, 33
Astragalus gummifer, 97
—— species, 100
Australian gums, 93

BALANOCARPUS, 6
Balsam, 38
Balsamea eyrthrea, 34
Balsamodendron, 33, 35
—— kafal, 71
Balsam of Copaiba, 38
—— ——, analytical values of, 39
—— of Peru, 46
—— ——, analytical values of, 47, 48
—— —— (artificial), 49
—— —— ——, analytical value of, 49
—— Tolu, 43

Bdellium, 33
——, analytical values of, 35
Beckerite, 18
Benjamin, 40
Benzaldehyde, 42
Benzoic acid, 19, 42, 43
Benzoin, 40
——, analytical values of, 42
Benzoresinol, 42
Benzyl benzoate, 43
—— cinnamate, 43
Beta amyrin, 9
Bisabol, 33
Blaze, 15
Boswellia carteri, 3
—— freriana, 73
—— freriana, 9
Botany Bay gum, 19
Boxed, 12
Breidine, 11
Brein, 11
Bryoidin, 11
Burgundy pitch, 60
Bursera gummifera, 10
Butea frondosa, 26
Button-lac, 27

CALLITRIS arenosa, 23
—— culcavata, 23
—— quadrivalis, 22
Calophyllum inophyllum, 77
Canada balsam, 58
—— ——, analytical values of, 58
—— ——, essential oil of, 59
Canarium, 8
—— luzonicum, 8, 9
—— zephyrenum, 9
Carre, 14
Castor, composition of, 70
Castor fiber, 69

INDEX

Castoreum, 69
Ceropetalum apetalum, gum from, 93
—— gummifera rum, gum from, 93
—— gums, analytical value of, 94
Chinese joss sticks, 34
Chir, 14
Chironium opoponax, 71
Cistus creticus, 74
—— cypricus, 74
—— ladaniferus, 74
Colophony, 7, 11
——, analytical values of, 16
Commiphora, 33, 35
Convolvulus scammonia, 65
Copaifera guibourtiana, 2
—— landsdorfii, 38
Copal, 1
——, analytical values of, 5, 6
Crude turpentine, 11
Cyanothryrsus ogea, 2,

DAEMONOROPS draco, 25
Dammar resin, 6
——, analytical values of, 7
Dammara Australis, 2
Daniella oblonga, 2
—— thurifera, 63
Devil's dung, 32
Dextropinene, 22
Dipentene, 11
Dipterocarpus, 40, 63
Diterpene, 24
Dorema ammoniacum, 35
Draco-alban, 25
Dragon's blood, 25
Driers, 16

ELAPHYRUM tomentosum, 77
Elemi, 8
——, analytical values of, 10
Elemic acid, 11
Essential oil distilled from Copaiba balsam, 39
Ester gums, 12

FERULA, 32
—— foetida, 32
—— narthex, 32
—— scorodorma, 32
—— tingitana, 36
Flindersia maculosa, gum from, 93
France, 14
Frankincense, 72

GALBANUM, 36
——, analytical values of, 37
Gamboge, 30
——, analytical values of, 32
Garcinia morella, 30
—— resin, 61
Garnet-lac, 27
Gedanite, 18
Ghatti gum, 90
—— ——, analytical values of, 92
Glessite, 18
Gommier resin, 10
Grass-tree gum, 19
Guaiacum, analytical values of, 24
—— officinale, 24
—— resin, 24
—— sanctum, 24
Gum Arabic, 78
—— ——, analyses of, 83
—— ——, analytical values of, 86
—— copal, 1
—— resins, 30
—— tragacanth, 97
Gurjun balsam, 63
—— ——, analytical values of, 64

HARDWICKIA manii, 63
Hashab gum, 87
Herabol myrrh, 33, 34
Hopea, 6
—— odorata, 7
Hymenoea species, 2

ICICA heptaphyllum, 77

INDEX

Illurin balsam, 63
—— ——, analytical values of, 63
Indian pine, 14
Insoluble gums, 101
Ipomoea orizabensis, 66
—— purga, 68

JALAP resin, 68
—— ——, analytical values of, 69

KAURI copal, 1
—— pine, 2

LADANUM, composition of, 76
——, essential oil of, 76
—— resin, 74
—— ——, analytical values of, 75
Larix decidua, 60
—— Europea, 60
Lead number, 33
Liquidambar styraciflua, 53

MACROZAMIA perowskiana, gum from, 93
—— spiralis, gum from, 93
Mastic resin, 23
——, analytic values of, 23, 24
Medicinal resins, 38
Melanorrhoea usitata, 62
Melia azadirachta gum of, 97
Myroxylon pereirae, 46
—— toluifera, 43
Myrrh, 33, 34
——, analytical values of, 35

OIL of elemi, 11
—— of turpentine, 11
Olibanum, 72
——, analytical values of, 71
Opopanax, 71
—— chironium, 71

PALEMBANG benzoin, 41
Palm dragon's blood, 25
Para-coumaric acid, 21

Persian insoluble gum, 95
Peucedanum galbaniflorum, 36
Phellandrene, 11
Phenyl-propyl cinnamate, 42
Picea excelsa, 60
Picric acid, 20
Pimaric acid, 22
Pinites succinifer, 19
Pinus Australis, 12
—— longifolia, 14
—— pinaster, 14
Pisticia lentiscus, 23
Podophyllin resin, 64
—— ——, analytical values of, 64
Podophyllotoxin, 64
Podophyllum emodi, 64
—— peltatum, 64
Protium heptaphyllum, 9
Pseudotsuga mucronata, 59

RESINS, Proper, 1
Retinodendron rassak, 8
Rhus vernicifera, 61
Rock dammar, 7, 8
—— ——, analytical values of 8
Rose dammar, 8
—— ——, analytical values of 8
Rosin oil, 17
Russia, 15

SAGAPENUM resin, 76
Sandarac, 22
Sanguis draconis, 25
Scammony resin, 65
—— ——, analytical values of 66, 67, 68
Sebonah, 72
Seedlac, 27
Senaar gum, 88
Senegal gum, 88
Shellac, 26
Shorea, 6
Siam benzoin, 40
Siaresinotannol, 42
Spirit varnishes, 6

Storax, 50
——, analytical values of, 53, 54, 56, 57
Storch-morawski reaction, 16, 24
Styracine, 42
Styrax, 40
—— benzoin, 40, 41
Styrol, 42
Succinite, 18
Sudan gum, 78
Sumatra benzoin, 41

TACAMAHAC resin, 77
Tachardia lacca, 26
Tateana, 21
T.N. shellac, 28
Toluresinotannol, 43
Trachylobium, 2

Trachylolic acid, 5
True gums, 78
Turpentine, 14, 15

UMBELLIFERONE, 76

VANILLIN, 42, 43
Venice turpentine, 60
—— ——, analytical values of, 61

WOOD oil, 63

XANTHORRHOEA, 19
—— arborea, 19, 22
—— Australis, 19, 22
—— drummondii, 21
—— hastilis, 19

THE END

AN ABRIDGED LIST OF THE
COMMERCIAL HANDBOOKS
OF
SIR ISAAC PITMAN & SONS, LTD.

[LONDON : 1 AMEN CORNER, E.C.4
BATH : Phonetic Institute. MELBOURNE : The Rialto, Collins St.
NEW YORK : 2 West 45th St.

The Prices contained in this Catalogue
:: apply only to the British Isles ::

TERMS—

Cash MUST *be sent with the order,* AND MUST INCLUDE AN APPROXIMATE AMOUNT FOR THE POSTAGE. *When a remittance is in excess of the sum required, the surplus will be returned.* Sums under 6d. *can be sent in stamps. For sums of* 6d. *and upwards Postal Orders or Money Orders are preferred to stamps, and should be crossed and made payable to*
SIR ISAAC PITMAN & SONS, LTD.
Remittances from abroad should be by means of International Money Orders in Foreign Countries, and by British Postal Orders within the British Overseas Dominions. Colonial Postal Orders are not negotiable in England. Foreign stamps CANNOT BE ACCEPTED.

ARITHMETIC

FIRST STEPS IN COMMERCIAL ARITHMETIC. By ARTHUR E. WILLIAMS, M.A., B.Sc. In crown 8vo, limp cloth, 80 pp. . . . . . **Net** 1/3
THE ELEMENTS OF COMMERCIAL ARITHMETIC. By THOMAS BROWN. In crown 8vo, cloth, 140 pp. . . . . . . **Net** 2/-
BUSINESS ARITHMETIC. Part I. In crown 8vo, cloth, 120 pp. 1/6. Answers 1/-
BUSINESS ARITHMETIC. Part II. In crown 8vo, cloth, 144 pp. 1/9. Answers 1/-
COMPLETE COMMERCIAL ARITHMETIC. Contains Parts I and II above mentioned. In crown 8vo, cloth, 264 pp. . . . . . . 3/-. Answers 1/6
SMALLER COMMERCIAL ARITHMETIC. By C. W. CROOK, B.A., B.Sc. In crown 8vo, cloth . . . . . . 1/6 net. Answers Net 1/6
FIRST STEPS IN WORKSHOP ARITHMETIC. By H. P. GREEN. In crown 8vo, limp cloth, about 80 pp. . . . . . . **Net** 1/3
COMPLETE MERCANTILE ARITHMETIC. With Elementary Mensuration. By H. P. GREEN, F.C.Sp.T. In crown 8vo, cloth gilt, with Key, 646 pp. . **Net** 5/-
Complete book without Key, 600 pp., 4/6 net. Key separately, 1/3 net. Also in three parts. Part I, 300 pp., 3/6 net. Part II, 208 pp., 1/6 net. Part III, 100 pp. 1/- net.
THE PRINCIPLES AND PRACTICE OF COMMERCIAL ARITHMETIC. By P. W. NORRIS, M.A., B.Sc. In demy 8vo, cloth, 452 pp. . . . **Net** 7/6
COUNTING HOUSE MATHEMATICS. By H. W. PORRITT and W. NICKLIN, A.S.A.A. In crown 8vo, cloth, 120 pp. . . . . . **Net** 1/6
ARITHMETIC AND BOOK-KEEPING. By THOS. BROWN, F.S.S., and VINCENT E. COLLINGE, A.C.I.S. In two parts. Each in crown 8vo, cloth. Part 1, 124 pp, 1/6 net. Part 2, 115 pp. . . . . . . **Net** 1/3
LOGARITHMS FOR BUSINESS PURPOSES. By H. W. PORRITT and W. NICKLIN, A.S.A.A. In crown 8vo, limp cloth . . . . . **Net** 9d.
RAPID METHODS IN ARITHMETIC. By JOHN JOHNSTON. Revised and Edited by G. K. BUCKNALL, A.C.I.S. (Hons.). New and Enlarged Edition. In foolscap 8vo, cloth, 96 pp. . . . . **Net** 1/-
EXERCISES ON RAPID METHODS IN ARITHMETIC. By JOHN JOHNSTON. In foolscap 8vo, cloth . . . . **Net** 1/-
METHOD IN ARITHMETIC. A guide to the teaching of Arithmetic. By G. R. PURDIE, B.A. In crown 8vo, cloth, 87 pp. . . . **Net** 1/6

THE METRIC AND BRITISH SYSTEM OF WEIGHTS, MEASURES, AND COINAGE. By Dr. F. MOLLWO PERKIN. In 8vo, with numerous illustrations.	Net	2/6
ARITHMETIC OF ALTERNATING CURRENTS. By E. H. CRAPPER, M.I.E.E. In crown 8vo, illustrated	Net	3/6
ARITHMETIC OF ELECTRICAL ENGINEERING. For Technical Students. In crown 8vo, illustrated	Net	3/6
THE SLIDE RULE: ITS OPERATIONS; AND DIGIT RULES. Pocket size	Net	6d.
THE SLIDE RULE. A Practical Manual. Illustrated	Net	3/6

BOOK-KEEPING AND ACCOUNTANCY

FIRST STEPS IN BOOK-KEEPING. By W. A. HATCHARD, A.C.P., F.B.T. In crown 8vo, limp cloth, 80 pp.	Net	1/3
PRIMER OF BOOK-KEEPING. Thoroughly prepares the student for the study of more elaborate treatises. In crown 8vo, cloth, 144 pp. 2/- Answers,	Net	1/6
EASY EXERCISES FOR PRIMER OF BOOK-KEEPING. In crown 8vo, 48 pp.		6d.
BOOK-KEEPING FOR BEGINNERS. A first course in the art of up-to-date Book-keeping. With Answers to the Exercises. By W. E. HOOPER, A.C.I.S. In crown 8vo, cloth, 148 pp.	Net	2/-
THE ELEMENTS OF BOOK-KEEPING. By W. O. BUXTON, A.C.A. (Hons.). In crown 8vo, cloth, 157 pp.	Net	2/-
BOOK-KEEPING AND COMMERCIAL PRACTICE. By H. H. SMITH, F.C.T., F.Inc.S.T. In crown 8vo, cloth, 152 pp.		1/9
BOOK-KEEPING SIMPLIFIED. Thoroughly revised edition. By W. O. BUXTON, A.C.A. (Hons.). In crown 8vo, cloth, 304 pp. 3/6. Answers,	Net	2/-
ADVANCED BOOK-KEEPING. In crown 8vo, cloth, 187 pp.	Net	3/6
HIGHER BOOK-KEEPING AND ACCOUNTS. By H. W. PORRITT and W. NICKLIN, A.S.A.A. In crown 8vo, cloth, 304 pp., with many up-to-date forms and facsimile documents	Net	3/6
FULL COURSE IN BOOK-KEEPING. By H. W. PORRITT and W. NICKLIN, A.S.A.A. In crown 8vo, cloth gilt, 540 pp.	Net	5/-
COMPLETE BOOK-KEEPING. A thoroughly comprehensive text-book, dealing with all departments of the subject. In crown 8vo, cloth, 424 pp. 5/- Answers,	Net	2/6
ADVANCED ACCOUNTS. A Practical Manual for the Advanced Student and Teacher. Edited by ROGER N. CARTER, M.Com., F.C.A. In demy 8vo, cloth gilt, 988 pp., with many forms and facsimile documents	Net	7/6
DICTIONARY OF BOOK-KEEPING. A Practical Guide and Book of Reference for Teachers, Students and Practitioners. By R. J. PORTERS. In demy 8vo, cloth gilt, with facsimiles, 780 pp.	Net	7/6
BOOK-KEEPING FOR RETAILERS. By H. W. PORRITT and W. NICKLIN, A.S.A.A. In crown 8vo, cloth, 124 pp.	Net	2/-
ADDITIONAL EXERCISES IN BOOK-KEEPING, Nos. I and II. New Editions. In crown 8vo, 56 pp. No. 1, 8d. net. No. 2, 9d. net Answers, each,	Net	6d.
BOOK-KEEPING TEST CARDS. Elem. and Inter. Per set	Net	1/6
BUSINESS BOOK-KEEPING. By J. ROUTLEY. In crown 8vo, cloth, 360 pp.	Net	3/6
EXAMINATION NOTES ON BOOK-KEEPING AND ACCOUNTANCY. By J. BLAKE HARROLD, A.C.I.S., F.C.R.A. Cloth, 6½ in. by 3½ in., 56 pp.		1/-
HOTEL BOOK-KEEPING. With illustrative forms and exercises. In crown 8vo, cloth, 72 pp.	Net	2/6
BOOK-KEEPING AND ACCOUNTANCY PROBLEMS. By G. JOHNSON, F.C.I.S. In crown 8vo, cloth gilt, 112 pp.	Net	2/6
COMBINED MANUSCRIPT BOOK FOR BOOK-KEEPING. In crown 4to, stiff paper wrapper, 96 pp.		1/6
IDEAL MANUSCRIPT BOOKS FOR BOOK-KEEPING. Specially ruled and adapted for working the exercises contained in the Primer of Book-keeping. The sets consist of: Cash Book; Purchase Book; Sales Book and Journal; Ledger. Each		6d.
AVON EXERCISE BOOKS FOR BOOK-KEEPING. Specially adapted for the exercises in *Book-keeping Simplified* or *Advanced Book-keeping*. Foolscap folio. Journal, 6d.; Cash Book, 6d.; Ledger		10d.
EXAMINATION NOTES ON MUNICIPAL ACCOUNTANCY. By W. G. DAVIS, A.S.A.A. Size 6½ in. by 3½ in., cloth, 56 pp.	Net	2/-
BALANCE SHEETS: HOW TO READ AND UNDERSTAND THEM. By PHILIP TOVEY, F.C.I.S. In foolscap 8vo, cloth, 85 pp., with 26 inset Balance Sheets	Net	2/-
HOW TO BECOME A QUALIFIED ACCOUNTANT. By R. A. WITTY, A.S.A.A. Second Edition. In crown 8vo, cloth, 120 pp.	Net	2/6
ACCOUNTANCY. By F. W. PIXLEY, F.C.A., *Barrister-at-Law*. In demy 8vo, cloth, 318 pp.	Net	6/-
BOOK-KEEPING TEACHERS' MANUAL. By C. H. KIRTON, A.C.I.S., F.Inc.S.T. In demy 8vo, cloth gilt, 230 pp.	Net	5/-
NOTES OF LESSONS ON BOOK-KEEPING. By J. ROUTLEY. In crown 8vo, 133 pp.	Net	3/-

AUDITING, ACCOUNTING AND BANKING. By Frank Dowler, A.C.A., and E. Mardinor Harris, A.I.B. In demy 8vo, cloth gilt, 328 pp. . . **Net**
PRINCIPLES OF BOOK-KEEPING EXPLAINED. By I. H. Humphrys. In crown 8vo, cloth, 120 pp. . . . . **Net**
MANUFACTURING BOOK-KEEPING AND COSTS. By G. Johnson, F.C.I.S. In demy 8vo, cloth gilt, 120 pp. . . . . . . **Net**
PRACTICAL BOOK-KEEPING. By the same Author. In demy 8vo, cloth, 420 pp. **Net**
DEPRECIATION AND WASTING ASSETS, and their Treatment in Computing Annual Profit and Loss. By P. D. Leake, F.C.A. In demy 8vo, cloth gilt, 257 pp. **Net**
THE PRINCIPLES OF AUDITING. By F. R. M. De Paula, F.C.A. In demy 8vo, cloth gilt, 224 pp. . . . . **Net**
COST ACCOUNTS IN PRINCIPLE AND PRACTICE. By A. Clifford Ridgway, F.C.A. In demy 8vo, cloth gilt, 120 pp. . . . . **Net**
GOLD MINE ACCOUNTS AND COSTING. A Practical Manual for Officials, Accountants, Book-keepers, Etc. By G. W. Tait. In demy 8vo, cloth gilt, 93 pp. . . . . **Net**
COMPANY ACCOUNTS. A complete, practical Manual for the use of officials in Limited Companies and advanced students. By Arthur Coles, F.C.I.S. In demy 8vo, cloth gilt, 356 pp. Second Edition . . . **Net**
THE ACCOUNTS OF EXECUTORS, ADMINISTRATORS AND TRUSTEES. By William B. Phillips, A.C.A. (Hons. Inter. and Final), A.C.I.S. In demy 8vo, cloth gilt, 152 pp. . . . . **Net**
RAILWAY ACCOUNTS AND FINANCE. The Railway Companies (Accounts and Returns) Act, 1911. By Allen E. Newhook, A.K.C. In demy 8vo, cloth gilt, 148 pp. . . . . **Net**
THE PERSONAL ACCOUNT BOOK. By W. G. Dousley, B.A. Size, 15¼ in. by 9½ in., half leather, 106 pp., with interleaved blotting paper . . **Net**
SHOPKEEPERS' ACCOUNTS SIMPLIFIED. By C. D. Cornell. In crown 8vo, 70 pp. . . . . **Net**
THE "EFFICIENT" CHECK FIGURE SYSTEM. By H. O. Horton. In demy 8vo. 22 pp. . . . . **Net**

BUSINESS TRAINING, COPY BOOKS, ETC.

COMMERCIAL READER (Junior Book). Our Food Supplies. By F. W. Chambers. With over 70 illustrations, 240 pp. . . . . **Net**
COMMERCIAL READER (Intermediate Book). Our Manufacturing Industries. In crown 8vo, cloth, 240 pp. Over 150 illustrations . . **Net**
COMMERCIAL READER (Senior Book). An introduction to Modern Commerce. Contains over 160 black and white illustrations. In crown 8vo, cloth, 272 pp. **Net**
OFFICE ROUTINE FOR BOYS AND GIRLS. In three stages. Each in crown 8vo, 64 pp. . . . . **Each**
FIRST STEPS IN BUSINESS TRAINING. By V. E. Collinge, A.C.I.S. In crown 8vo, limp cloth, 80 pp. . . . . . . **Net**
COUNTING-HOUSE ROUTINE. 1st Year's Course. By Vincent E. Collinge, A.C.I.S. In crown 8vo, cloth, with illustrations, maps, and facsimile commercial forms, 162 pp. . . . . **Net**
COUNTING-HOUSE ROUTINE. 2nd Year's Course. By Vincent E. Collinge, A.C.I.S. In crown 8vo, cloth, with illustrations, maps and facsimile commercial forms, 188 pp. . . . . **Net**
THE PRINCIPLES OF BUSINESS. By James Stephenson, M.A., M.Com., B.Sc.
Part 1. In crown 8vo, cloth, 217 pp. . . . . **Net**
Part 2. In crown 8vo, cloth, 320 pp. . . . . **Net**
MANUAL OF BUSINESS TRAINING. Contains 66 maps and facsimiles. Eighth Edition, thoroughly revised and considerably enlarged. In crown 8vo, cloth, 302 pp. . . . .
THE PRINCIPLES AND PRACTICE OF COMMERCE. By James Stephenson, M.A., M.Com., B.Sc. In demy 8vo, cloth gilt, 648 pp., with many illustrations, diagrams, etc. . . . . **Net**
COMMERCIAL PRACTICE. By Alfred Schofield. In crown 8vo, cloth, 296 pp. . . . . **Net**
THE THEORY AND PRACTICE OF COMMERCE. Being a Complete Guide to Methods and Machinery of Business. Edited by F. Heelis, F.C.I.S., Assisted by Specialist Contributors. In demy 8vo, cloth gilt, 620 pp., with many facsimile forms 6/- Net. Also in 2 vols., each . . . . **Net**
HOW TO TEACH BUSINESS TRAINING. By F. Heelis, F.C.I.S. In crown 8vo, 160 pp. . . . . **Net**
QUESTIONS IN BUSINESS TRAINING. By F. Heelis, F.C.I.S. In crown 8vo, cloth, 108 pp. . . . . **Net**

MODERN BUSINESS AND ITS METHODS. By W. CAMPBELL, Chartered Secretary. In two vols. Each **3/6 Net.** Complete . . . . . **Net** 6/-
A COURSE IN BUSINESS TRAINING. By G. K. BUCKNALL, A.C.I.S. In crown 8vo, 192 pp. 2/6
FACSIMILE COMMERCIAL FORMS. New, Revised, and Enlarged Edition. Thirty-five separate forms in envelope . . . . . **Net** 1/-
    Forms separately, per doz. . . . . . . **Net** 4d.
EXERCISE BOOK OF FACSIMILE COMMERCIAL FORMS. In large post 4to, 32 pp. **Net** 8d.
FACSIMILE COMPANY FORMS. Thirty-four separate forms in envelope . **Net** 1/3
    Forms separately, per doz. . . . . . . **Net** 6d.
"NEW ERA" BUSINESS COPY BOOKS. By F. HEELIS, F.C.I.S. Civil Service Style. In three books, Junior, Intermediate, and Senior. Each in stout paper covers, large post 4to, 32 pp. . . . . . **Net** 6d.
BUSINESS TRAINING EXERCISE BOOK. Part 1. By JAMES E. SLADEN, M.A. (Oxon.), F.I.S.A. In large post 4to, 64 pp. . . . **Net** 8d.
MANUSCRIPT LETTERS AND EXERCISES. In envelope. . . **Net** 8d.
OFFICE ROUTINE COPY BOOKS, No. 1, 8d. net, No. 2, 6d. and No. 3. 6d. Each in large post 4to, 24 pp.
COMMERCIAL HANDWRITING AND CORRESPONDENCE. In foolscap 4to, quarter cloth, 80 pp. . . . . . . **Net** 2/-
BUSINESS HANDWRITING. Seventh Edition, Revised. In crown 8vo, cloth, 84 pp. . . . . . . **Net** 1/6
HOW TO WRITE A GOOD HAND. By B. T. B. HOLLINGS. In crown 8vo, oblong, 56 pp. . . . . . . **Net** 1/6
HANDBOOK FOR COMMERCIAL TEACHERS. By FRED HALL, M.A., B.Com., F.C.I.S., etc. In crown 8vo, cloth gilt, 200 pp. . **Net** 2/6
THE BUSINESS GIRL'S HANDBOOK. By C. CHISHOLM, M.A., and D. W. WALTON. Foreword by SARAH BERNHARDT. In crown 8vo, cloth, 176 pp. . **Net** 2/6
THE BOY'S BOOK OF BUSINESS. By the same Authors. Foreword by Lieut.-Gen. Sir R. S. S. BADEN-POWELL. In crown 8vo, cloth, 176 pp. . **Net** 2/-
BUSINESS METHODS AND SECRETARIAL WORK FOR GIRLS AND WOMEN. By HELEN REYNARD, M.A. In crown 8vo, cloth, 96 pp. . **Net** 1/6
THE JUNIOR WOMAN SECRETARY. By ANNIE E. DAVIS, F.Inc.S.T. In crown 8vo, cloth, 100 pp., with illustrations . . **Net** 1/3
THE JUNIOR CORPORATION CLERK. By J. B. CARRINGTON, F.S.A.A. In crown 8vo, cloth gilt, with illustrations, 136 pp. . . **Net** 1/6
POPULAR GUIDE TO JOURNALISM. By A. KINGSTON. 4th Edition. In crown 8vo, 124 pp., cloth . . . . . **Net** 1/6
PRACTICAL JOURNALISM AND NEWSPAPER LAW. By A. BAKER, M.J.I., and E. A. COPE. In crown 8vo, cloth, 180 pp. . . **Net** 3/6

CIVIL SERVICE

CIVIL SERVICE GUIDE. By A. J. LAWFORD JONES. In crown 8vo, cloth, 129 pp. **Net** 1/6
DIGESTING RETURNS INTO SUMMARIES. By A. J. LAWFORD JONES, of H.M. Civil Service. In crown 8vo, cloth, 84 pp. . . **Net** 2/-
COPYING MANUSCRIPT, ORTHOGRAPHY, HANDWRITING, etc. By the same Author. Actual Examination Papers only. In foolscap folio, 48 pp. . **Net** 2/-
CIVIL SERVICE HANDWRITING GUIDE AND COPY BOOK. By H. T. JESSOP, B.Sc. In crown 4to, 32 pp. . . . . **Net** 6d.
CIVIL SERVICE AND COMMERCIAL COPYING FORMS. In crown 8vo, 40 pp. **Net** 6d.
RULED FORMS FOR USE WITH THE ABOVE. Books I and II. Each foolscap folio, 40 pp. . . . . . . **Net** 8d.
CIVIL SERVICE AND COMMERCIAL LONG AND CROSS TOTS. Two Series, each in crown 8vo, 48 pp. . . . . . **Net** 6d.
CIVIL SERVICE ARITHMETIC TESTS. By P. J. VARLEY-TIPTON. In crown 8vo, cloth, 102 pp. . . . . . **Net** 2/-
CIVIL SERVICE ESSAY WRITING. By W. J. ADDIS, M.A. In crown 8vo, limp cloth, 108 pp. . . . . . **Net** 2/-
CIVIL SERVICE PRACTICE IN PRÉCIS WRITING. Edited by ARTHUR REYNOLDS, M.A. (Oxon.). In crown 8vo, cloth, 240 pp. . . **Net** 2/6
ELEMENTARY PRÉCIS WRITING. By WALTER SHAWCROSS, B.A. In crown 8vo, cloth, 80 pp. . . . . . **Net** 1/3
GUIDE TO INDEXING AND PRÉCIS WRITING. By W. J. WESTON, M.A., B.Sc. (Lond.), and E. BOWKER. In crown 8vo, cloth, 110 pp. . **Net** 1/6
INDEXING AND PRÉCIS WRITING. By A. J. LAWFORD JONES. In crown 8vo, cloth, 144 pp. . . . . . **Net** 2/-
EXERCISES AND ANSWERS IN INDEXING AND PRÉCIS WRITING. By W. J. WESTON, M.A., B.Sc. (Lond.). In crown 8vo, cloth, 144 pp. . **Net** 2/-

ENGLISH AND COMMERCIAL CORRESPONDENCE

FIRST STEPS IN COMMERCIAL ENGLISH. By W. J. WESTON, M.A., B.Sc. (Lond.). In crown 8vo, limp cloth, 80 pp. . . . . . . Net 1/3
FIRST STEPS IN BUSINESS LETTER WRITING. By FRED HALL, M.A., B.Com., F.C.I.S., etc. In crown 8vo, limp cloth, 80 pp. . . . . Net 1/3
GUIDE TO COMMERCIAL CORRESPONDENCE AND BUSINESS COMPOSITION. By W. J. WESTON, M.A., B.Sc. (Lond.). In crown 8vo, cloth, 156 pp., with many facsimile commercial documents 2/6
MANUAL OF COMMERCIAL ENGLISH. By WALTER SHAWCROSS, B.A. Including Composition and Précis Writing. In crown 8vo, cloth gilt, 234 pp. . Net 2/6
HOW TO TEACH COMMERCIAL ENGLISH. By WALTER SHAWCROSS, B.A. In crown 8vo, cloth gilt, 160 pp. . . . . Net 3/6
COMMERCIAL CORRESPONDENCE AND COMMERCIAL ENGLISH. In crown 8vo, cloth, 272 pp. 3/6
PRINCIPLES AND PRACTICE OF COMMERCIAL CORRESPONDENCE. By J. STEPHENSON, M.A., M.Com., B.Sc. In demy 8vo, 320 pp. . . Net 7/6
ENGLISH MERCANTILE CORRESPONDENCE. In crown 8vo, cloth gilt, 260 pp. Net 3/6
FIRST STEPS IN BUSINESS COMPOSITION. Edited by R. W. HOLLAND, M.A., M.Sc., LL.D. In crown 8vo, limp cloth, 80 pp. . . . . Net 1/3
ENGLISH COMPOSITION AND CORRESPONDENCE. By J. F. DAVIS, D.Lit., M.A., LL.B. (Lond.). In crown 8vo, cloth, 118 pp. . . . . . Net 2/-
A GUIDE TO ENGLISH COMPOSITION. By the Rev. J. H. BACON. 112 pp. cloth Net 2/-
ENGLISH GRAMMAR. New Edition, Revised and Enlarged by C. D. PUNCHARD, B.A. (Lond.). In crown 8vo, cloth, 142 pp. . . . . Net 2/-
ENGLISH GRAMMAR AND COMPOSITION. By W. J. WESTON, M.A., B.Sc. (Lond.). In crown 8vo, cloth, 320 pp. . . . . . Net 3/6
ENGLISH PROSE COMPOSITION. By W. J. WESTON, M.A., B.Sc. In crown 8vo, cloth, 224 pp. . . . . Net 3/6
SELF-HELP EXERCISES IN ENGLISH (Reform Method). In crown 8vo, limp cloth, 80 pp. . . . . Net 1/3
NOTES OF LESSONS ON ENGLISH. In crown 8vo, cloth, 208 pp. . Net 3/6
PUNCTUATION AS A MEANS OF EXPRESSION. By A. E. LOVELL, M.A. In crown 8vo, cloth, 80 pp. . . . . Net 1/-
PRÉCIS WRITING } (See CIVIL SERVICE, page 4)
ESSAY WRITING
STUDIES IN ELOCUTION. By E. M. CORBOULD (*Mrs. Mark Robinson*). With over 100 selections for Reciters and Readers. In crown 8vo, cloth gilt, 270 pp Net 3/6
POCKET DICTIONARY. Royal 32mo, 5 in. by 3 in., cloth gilt, 363 pp. . Net 1/6
COMMERCIAL DICTIONARY. In foolscap 8vo, paper boards, 192 pp. . Net 1/-
BOOK OF HOMONYMS. With copious Exercises on Homogeneous, and Homophonous Words and chapters on Compound Hyphenated Words, etc. By B. S. BARRETT. In crown 8vo, cloth, 203 pp. . . . . . Net 2/-

COMMERCIAL GEOGRAPHY

FIRST STEPS IN COMMERCIAL GEOGRAPHY. By JAMES STEPHENSON, M.A., B.Com. There are 16 maps and diagrams included. In crown 8vo, limp cloth, 80 pp. . . . . Net 1/3
THE WORLD AND ITS COMMERCE. In crown 8vo, cloth, 128 pp., with 34 maps . 1/3
THE ELEMENTS OF COMMERCIAL GEOGRAPHY. By C. H. GRANT, M.Sc., F.R.Met.Soc. In crown 8vo, cloth, 140 pp. . . . . . Net 2/-
COMMERCIAL GEOGRAPHY OF THE BRITISH ISLES. In crown 8vo, cloth, 150 pp., with 34 coloured maps and plates, three black and white maps, and other illustrations . . . . Net 2/-
COMMERCIAL GEOGRAPHY OF THE BRITISH EMPIRE ABROAD AND FOREIGN COUNTRIES. In crown 8vo, cloth, 205 pp., with 35 coloured maps and plates, 11 black and white maps, and end-paper maps . . Net 2/-
COMMERCIAL GEOGRAPHY OF THE WORLD. In crown 8vo, cloth, 350 pp., with about 90 maps and plates . . . . Net 4/-
EXAMINATION NOTES ON COMMERCIAL GEOGRAPHY. By W. P. RUTTER, M.Com. Size 6¼ in. by 3¾ in., cloth, 120 pp. . . . . Net 1/-
ECONOMIC GEOGRAPHY. (*See* "ECONOMICS" page 6)
THE ELEMENTS OF COMMERCIAL HISTORY. By FRED HALL, M.A., B.Com., F.C.I.S. In crown 8vo, cloth, 164 pp. . . . . . Net 2/-

COMMERCIAL HISTORY

COMMERCIAL HISTORY. By J. R. V. MARCHANT, M.A. In crown 8vo, cloth gilt, 272 pp. . . . . **Net** 4/6
PRINCIPLES OF COMMERCIAL HISTORY. By J. STEPHENSON, M.A., M.Com., B.Sc. In demy 8vo, cloth, 279 pp. . . . . **Net** 7/6
ECONOMIC HISTORY. (*See* "ECONOMICS" below.)

ECONOMICS

THE ELEMENTS OF POLITICAL ECONOMY. By H. HALL, B.A. In crown 8vo, cloth, 140 pp. . . . . **Net** 2/-
GUIDE TO POLITICAL ECONOMY. By F. H. SPENCER, D.Sc., LL.B. In crown 8vo, cloth gilt, 232 pp. . . . . **Net** 3/6
OUTLINES OF THE ECONOMIC HISTORY OF ENGLAND: A Study in Social Development. By H. O. MEREDITH, M.A., M.Com. In demy 8vo, cloth gilt, 376 pp. . . . . **Net** 6/-
ECONOMIC GEOGRAPHY. By JOHN McFARLANE, M.A., M.Com. In demy 8vo, cloth gilt, 568 pp., 18 illustrations . . . . . . **Net** 8/6
THE HISTORY AND ECONOMICS OF TRANSPORT. By A. W. KIRKALDY, M.A., B.Litt. (Oxford), M.Com. (Birm.), and A. DUDLEY EVANS. In demy 8vo, cloth gilt, 350 pp. . . . . **Net** 7/6
THE ECONOMICS OF TELEGRAPHS AND TELEPHONES. By JOHN LEE, M.A. In crown 8vo, cloth gilt, 92 pp. . . . . . . **Net** 2/6
INDUSTRY AND FINANCE. (Supplementary Volume.) Edited by A. W. KIRKALDY, M.A., B.Litt., M.Com. In demy 8vo, cloth, 180 pp. . . . **Net** 5/-
OUTLINES OF LOCAL GOVERNMENT. By JOHN J. CLARKE, M.A., F.S.S. In crown 8vo, 83 pp. . . . . **Net** 1/6
OUTLINES OF CENTRAL GOVERNMENT. By the same Author. In crown 8vo. 90 pp. . . . . **Net** 1/6
OUTLINES OF INDUSTRIAL AND SOCIAL ECONOMICS. By the same Author. In crown 8vo, 108 pp. . . . . **Net** 1/6

BANKING AND FINANCE

THE ELEMENTS OF BANKING. By J. P. GANDY. In crown 8vo, cloth, 140 pp. **Net** 2/-
BANK ORGANIZATION, MANAGEMENT, AND ACCOUNTS. By J. F. DAVIS, M.A., D.Lit., LL.B. (Lond.) In demy 8vo, cloth gilt, 165 pp., with forms . . **Net** 5/-
MONEY, EXCHANGE, AND BANKING. In their Practical, Theoretical, and Legal Aspects. By H. T. EASTON, A.I.B. Second Edition, Revised. In demy 8vo, cloth, 312 pp. . . . . **Net** 6/-
PRACTICAL BANKING. By J. F. G. BAGSHAW. With Chapters on **The Principles of Currency**, by C. F. HANNAFORD, A.I.B., and **Bank Book-keeping**, by W. H. PEARD. In demy 8vo, cloth gilt, about 400 pp. . . . . **Net** 6/-
BANKERS' SECURITIES AGAINST ADVANCES. By LAWRENCE A. FOGG, Cert. A.I.B. In demy 8vo, cloth gilt, 123 pp. . . . . **Net** 5/-
BANKERS' ADVANCES. By F. R. STEAD. Edited by Sir JOHN PAGET, K.C. In demy 8vo, cloth, 144 pp. . . . . **Net** 6/-
FOREIGN EXCHANGE, A PRIMER OF. By W. F. SPALDING. In crown 8vo., cloth, 108 pp. . . . . **Net** 3/6
FOREIGN EXCHANGE AND FOREIGN BILLS IN THEORY AND IN PRACTICE. By W. F. SPALDING, Cert. A.I.B. In demy 8vo, cloth gilt, 227 pp. . . **Net** 7/6
EASTERN EXCHANGE. By W. F. SPALDING. In demy 8vo, cloth, 375 pp., illustrated . . . . **Net** 12/6
TALKS ON BANKING TO BANK CLERKS. By H. E. EVANS. In crown 8vo, cloth . . . . **Net** 2/6
SIMPLE INTEREST TABLES. By Sir WILLIAM SCHOOLING, K.B.E. In crown 4to, cloth gilt . . . . **Net** 21/-
DICTIONARY OF BANKING. A Complete Encyclopaedia of Banking Law and Practice. By W. THOMSON and LLOYD CHRISTIAN. Third Edition. In crown 4to, half leather gilt, 641 pp. . . . . . . **Net** 30/-
NOTES ON BANKING AND COMMERCIAL LAW. By T. LLOYD DAVIES. In f'cap 8vo, 100 pp. . . . . **Net** 3/-

INSURANCE

THE ELEMENTS OF INSURANCE. By J. ALFRED EKE. In crown 8vo, cloth, 140 pp. . . . . **Net 2/**
INSURANCE. By T. E. YOUNG, B.A., F.R.A.S. A complete and practical exposition. With sections on Workmen's Compensation Insurance, by W. R. STRONG, F.I.A., and The National Insurance Scheme, by VYVYAN MARR, F.F.A., F.I.A. Third Edition. Revised and Enlarged. In demy 8vo, cloth gilt, 440 pp. . **Net 10/**
GUIDE TO LIFE ASSURANCE. By S. G. LEIGH, F.I.A. In crown 8vo, cloth gilt, 192 pp. . . . . **Net 3/**
INSURANCE OFFICE ORGANIZATION, MANAGEMENT, AND ACCOUNTS. By T. E. YOUNG, B.A., F.R.A.S., and RICHARD MASTERS, A.C.A. Second Edition, Revised. In demy 8vo, cloth gilt, 146 pp. . . . . **Net 5/**
GUIDE TO MARINE INSURANCE. By HENRY KEATE. In crown 8vo, cloth gilt, 203 pp. . . . . **Net 3/**
THE PRINCIPLES OF MARINE LAW. (*See* p. 10.)

SHIPPING

SHIPPING. By A. HALL and F. HEYWOOD. In crown 8vo, cloth, 136 pp. . **Net 2/**
SHIPPING OFFICE ORGANIZATION, MANAGEMENT, AND ACCOUNTS. By ALFRED CALVERT. In demy 8vo, cloth gilt, 203 pp. . . . **Net 6/**
THE EXPORTER'S HANDBOOK AND GLOSSARY. By F. M. DUDENEY. With Foreword by W. EGLINGTON. In demy 8vo, cloth gilt, 254 pp. . **Net 6/—**
CONSULAR REQUIREMENTS FOR EXPORTERS AND SHIPPERS TO ALL PARTS OF THE WORLD. By J. S. NOWERY. In crown 8vo, cloth, 82 pp. . **Net 2/**
CASE AND FREIGHT COSTS. The principles of calculation relating to the cost of, and freight on, sea or commercial cases. By A. W. E. CROSFIELD. In crown 8vo, cloth, 62 pp. . . . . **Net 2/—**
HOW TO DO BUSINESS WITH RUSSIA. By C. E. W. PETERSSON and W. BARNES STEVENI. In demy 8vo, cloth, 200 pp. . . . . . **Net 5/**

SECRETARIAL WORK

COMPANY SECRETARIAL WORK. By E. MARTIN, F.C.I.S. In crown 8vo, cloth, 154 pp. . . . . **Net 2/-**
GUIDE TO COMPANY SECRETARIAL WORK. By O. OLDHAM, A.C.I.S. In crown 8vo, cloth gilt, 256 pp. . . . . **Net 3/**
GUIDE FOR THE COMPANY SECRETARY. By ARTHUR COLES, F.C.I.S. Illustrated with 76 facsimile forms. Second Edition, Revised and Enlarged. In demy 8vo, cloth gilt, 432 pp. . . . . **Net 6/-**
SECRETARY'S HANDBOOK. Edited by HERBERT E. BLAIN. In demy 8vo, cloth gilt, 168 pp. . . . . **Net 5/-**
THE CHAIRMAN'S MANUAL. By GURDON PALIN, *of Gray's Inn, Barrister-at-Law,* and ERNEST MARTIN, F.C.I.S. In crown 8vo, cloth gilt, 192 pp. . **Net 3/**
PROSPECTUSES: HOW TO READ AND UNDERSTAND THEM. By PHILIP TOVEY, F.C.I.S. In demy 8vo, cloth gilt, 109 pp. . . . **Net 2/**
OUTLINES OF TRANSFER PROCEDURE IN CONNECTION WITH STOCKS, SHARES, AND DEBENTURES OF JOINT STOCK COMPANIES. By F. D. HEAD, B.A. (Oxon), *of Lincoln's Inn, Barrister-at-Law.* In demy 8vo, cloth gilt, 112 pp. . . . . **Net 2/**
WHAT IS THE VALUE OF A SHARE? By D. W. ROSSITER. In demy 8vo, limp cloth, 20 pp. . . . . **Net 2/**
HOW TO TAKE MINUTES. Edited by E. MARTIN, F.C.I.S. Second Edition, Enlarged and Revised. In demy 8vo, cloth, 126 pp. . . **Net 2/**
DICTIONARY OF SECRETARIAL LAW AND PRACTICE. A comprehensive Encyclopaedia of information and direction on all matters connected with the work of a Company Secretary. Fully illustrated with the necessary forms and documents. With sections on special branches of Secretarial Work. With contributions by nearly 40 eminent authorities. Edited by PHILIP TOVEY, F.C.I.S. In one vol., half leather gilt, 1011 pp. Third Edition, Revised and Enlarged . **Net 42/-**
FACSIMILE COMPANY FORMS. (*See* p. 4.)
COMPANY ACCOUNTS. (*See* p. 3.)
COMPANY LAW. (*See* p. 11.)

INCOME TAX

PRACTICAL INCOME TAX. A Guide to the Preparation of Income Tax Returns. By W. E. SNELLING. In crown 8vo, cloth, 136 pp. . . **Net 2/6**

INCOME TAX AND SUPER-TAX PRACTICE. Including a Dictionary of Income Tax and specimen returns, showing the effect of recent enactments down to the Finance Act, 1918, and Decisions in the Courts. By W. E. SNELLING. Third Edition, Revised and Enlarged. In demy 8vo, cloth gilt, 518 pp. . . **Net** 15/-
COAL MINES EXCESS PAYMENTS. Guarantee Payments and Levies for Closed Mines. By W. E. SNELLING. In demy 8vo, cloth gilt, 180 pp. . . . **Net** 12/6
INCOME TAX AND SUPER-TAX LAW AND CASES. Including the Finance Act, 1918. By W. E. SNELLING. Third Edition, Revised. In demy 8vo, cloth gilt, 472 pp . . . . **Net** 12/6
EXCESS PROFITS (including Excess Mineral Rights) **DUTY,** and Levies under the Munitions of War Acts. By W. E. SNELLING. Fifth Edition, Revised and Enlarged. In demy 8vo, cloth gilt, 422 pp. . . . . . **Net** 15/-

BUSINESS ORGANIZATION AND MANAGEMENT

THE PSYCHOLOGY OF MANAGEMENT. By L. M. GILBRETH. In demy 8vo, cloth gilt, 354 pp. . . . . **Net** 7/6
EMPLOYMENT MANAGEMENT. Compiled and edited by DANIEL BLOOMFIELD. In demy 8vo, cloth, 507 pp. . . . . **Net** 8/6
OFFICE ORGANIZATION AND MANAGEMENT, INCLUDING SECRETARIAL WORK. By LAWRENCE R. DICKSEE, M.Com., F.C.A., and H. E. BLAIN. Fourth Edition, Revised. In demy 8vo, cloth gilt, 314 pp. . . **Net** 7/6
MUNICIPAL ORGANIZATION AND MANAGEMENT. Edited by W. BATESON, A.C.A., F.S.A.A., In crown 4to, half leather gilt, with 250 forms, diagrams, etc., 503 pp. . . . . **Net** 25/-
COUNTING-HOUSE AND FACTORY ORGANIZATION. By J. GILMOUR WILLIAMSON. In demy 8vo, cloth gilt, 182 pp. . . . . . . **Net** 6/-
SOLICITORS' OFFICE ORGANIZATION, MANAGEMENT, AND ACCOUNTS. By E. A. COPE, and H. W. H. ROBINS. In demy 8vo, cloth gilt, 176 pp., with numerous forms . . . . **Net** 5/-
COLLIERY OFFICE ORGANIZATION AND ACCOUNTS. By J. W. INNES, F.C.A., and T. COLIN CAMPBELL, F.C.I. In demy 8vo, cloth gilt, 135 pp. . **Net** 6/-
CLUBS AND THEIR MANAGEMENT. By FRANCIS W. PIXLEY, F.C.A. *Of the Middle Temple, Barrister-at-Law.* In demy 8vo, cloth gilt, 240 pp. . **Net** 7/6
DRAPERY BUSINESS ORGANIZATION, MANAGEMENT AND ACCOUNTS. By J. ERNEST BAYLEY. In demy 8vo, cloth gilt, 302 pp. . . . **Net** 7/6
GROCERY BUSINESS ORGANIZATION AND MANAGEMENT. By C. L. T. BEECHING and J. ARTHUR SMART. Second Edition. In demy 8vo, cloth, 160 pp. . . . . **Net** 6/-
INDUSTRIAL TRAFFIC MANAGEMENT. By GEO. B. LISSENDEN. With a Foreword by C. E. MUSGRAVE. In demy 8vo, cloth gilt, 260 pp. . **Net** 7/6
SHIPPING ORGANIZATION, MANAGEMENT AND ACCOUNTS. (*See p. 7.*)
INSURANCE OFFICE ORGANIZATION, MANAGEMENT AND ACCOUNTS. (*See p. 7.*)
BANK ORGANIZATION AND MANAGEMENT. (*See p. 6.*)
THE CARD INDEX SYSTEM. In crown 8vo, 100 pp. . . . **Net** 2/-
FILING SYSTEMS. By E. A. COPE. In crown 8vo, cloth gilt, 200 pp., **Net** 2/6
A MANUAL OF DUPLICATING METHODS. By W. DESBOROUGH. In demy 8vo, cloth, 90 pp. . . . . **Net** 2/-

ADVERTISING AND SALESMANSHIP

ADVERTISING. By HOWARD BRIDGEWATER. In crown 8vo, cloth, 100 pp. . **Net** 2/-
ADS. AND SALES. By HERBERT N. CASSON. In demy 8vo, cloth, 167 pp., . **Net** 7/6
THE THEORY AND PRACTICE OF ADVERTISING. By W. DILL SCOTT, Ph.D. In large crown 8vo, cloth, 61 illustrations . . . . . **Net** 7/6
ADVERTISING AS A BUSINESS FORCE. By P. T. CHERINGTON. In demy 8vo, cloth gilt, 586 pp. . . . . **Net** 7/6
THE PRINCIPLES OF ADVERTISING ARRANGEMENT. By F. A. PARSONS. Size 7 in. by 10¼ in., cloth, 128 pp., with many illustrations . . **Net** 7/6
THE NEW BUSINESS. By HARRY TIPPER. In demy 8vo, cloth gilt, 406 pp. **Net** 8/6
THE CRAFT OF SILENT SALESMANSHIP. A Guide to Advertisement Construction. By C. MAXWELL TREGURTHA and J. W. FRINGS. Foreword by T. SWINBORNE SHELDRAKE. Size 6¼ in. by 9¼ in., cloth, 98 pp., with illustrations . **Net** 5/-
THE PSYCHOLOGY OF ADVERTISING. By W. DILL SCOTT, Ph.D. In demy 8vo, with 67 illustrations . . . . **Net** 7/6
HOW TO ADVERTISE. By G. FRENCH. In crown 8vo, cloth, with many illustrations . . . . **Net** 8/6
THE MANUAL OF SUCCESSFUL STOREKEEPING. By W. R. HOTCHKIN. In demy 8vo, cloth, 298 pp. . . . . **Net** 8/6

SALESMANSHIP. By W. A. GORBION and G. E. GRIMSDALE. In crown 8vo, cloth, 186 pp. . . . . **Net** 2/(

PRACTICAL SALESMANSHIP. By N. C. FOWLER, assisted by 29 expert Salesmen, etc. In crown 8vo, cloth, 337 pp. . . . . **Net** 5/-

COMMERCIAL TRAVELLING. By ALBERT E. BULL. In crown 8vo, cloth gilt, 170 pp. . . . . **Net** 3/(

BUSINESS HANDBOOKS AND WORKS OF REFERENCE

COMMERCIAL ENCYCLOPAEDIA AND DICTIONARY OF BUSINESS. Edited by J. A. SLATER, B.A., LL.B. (Lond.), *Barrister-at-Law*. Assisted by about 50 specialists as contributors. A reliable and comprehensive work of reference on all commercial subjects, specially written for the busy merchant, the commercial student, and the modern man of affairs. With numerous maps, illustrations, facsimile business forms and legal documents, diagrams, etc. In 4 vols., large crown 4to (each about 450 pp.), cloth gilt . . . . . . **Net** £5
Half leather gilt . . . . **Net** £2 12s. 6d

COMMERCIAL SELF-EDUCATOR. A comprehensive guide to business specially designed for commercial students, clerks, and teachers. Edited by ROBERT W. HOLLAND, M.A., M.Sc., LL.D. Assisted by upwards of 40 Specialists as contributors. With many maps, illustrations, documents, Diagrams, etc. Complete in 2 vols., crown 4to, cloth gilt, about 980 pp., sprinkled edges . . **Net** 18/-

BUSINESS MAN'S GUIDE. Edited by J. A. SLATER, B.A., LL.B. Seventh Edition, Revised. In crown 8vo, cloth, 520 pp. . . . . **Net** 5/-

COMMERCIAL ARBITRATIONS. By E. J. PARRY, B.Sc., F.I.C., F.C.S. In crown 8vo, cloth gilt, 105 pp. . . . . **Net** 3/6

LECTURES ON BRITISH COMMERCE, INCLUDING FINANCE, INSURANCE, BUSINESS AND INDUSTRY. By the Rt. Hon. FREDERICK HUTH JACKSON, G. ARMITAGE-SMITH, M.A., D.Litt., ROBERT BRUCE, C.B., etc. In demy 8vo, cloth gilt, 295 pp. . . . . **Net** 7/6

THE MONEY AND THE STOCK AND SHARE MARKETS. By EMIL DAVIES. In crown 8vo, cloth, 124 pp. . . . . **Net** 2/-

THE EVOLUTION OF THE MONEY MARKET (1385-1915). By ELLIS T. POWELL, LL.B. (Lond.), D.Sc. (Econ.) (Lond.) In demy 8vo, cloth gilt, 748 pp. . **Net** 10/6

THE HISTORY, LAW, AND PRACTICE OF THE STOCK EXCHANGE. By A. P. POLEY, B.A., *Barrister-at-Law*, and F. H. CARRUTHERS GOULD, *of the Stock Exchange*. Third Edition, Revised. In demy 8vo, cloth gilt, 348 pp. . **Net** 7/6

DICTIONARY OF THE WORLD'S COMMERCIAL PRODUCTS. By J. A. SLATER, B.A., LL.B. (Lond.). Second Edition. In demy 8vo, cloth, 170 pp. . **Net** 3/6

TELEGRAPH CIPHERS. A condensed vocabulary of 101,000,000 pronounceable artificial words, all of ten letters. By A. W. E. CROSFIELD. Size 12 in. by 12 in., cloth . . . . **Net** 21/-

DISCOUNT, COMMISSION, AND BROKERAGE TABLES. By ERNEST HEAVINGHAM. Size 3 in. by 4¼ in., cloth, 160 pp. . . . . **Net** 1/6

BUSINESS TERMS, PHRASES, AND ABBREVIATIONS. Fourth Edition, Revised and Enlarged. In crown 8vo, cloth, 280 pp. . . . . . **Net** 3/-

MERCANTILE TERMS AND ABBREVIATIONS. Containing over 1,000 terms and 500 abbreviations used in commerce, with definitions. Size 3 in. by 4¼ in., cloth, 126 pp. . . . . **Net** 1/6

TRAMWAY RATING VALUATIONS AND INCOME TAX ASSESSMENTS. By F. A. MITCHESON. In demy 8vo, cloth gilt . . . . . **Net** 2/6

THE TRADER'S GUIDE TO COUNTY COURT PROCEDURE. In foolscap 8vo, cloth, 112 pp. . . . . **Net** 1/6

A COMPLETE GUIDE TO THE IMPROVEMENT OF THE MEMORY. By the late Rev. J. H. BACON. In foolscap 8vo, cloth, 118 pp. . . . . **Net** 1/6

HOW TO STUDY AND REMEMBER. By B. J. DAVIES. Third Edition. In crown 8vo . . . . **Net** 9d.

TRADER'S HANDBOOKS. In crown 8vo, cloth, 260 pp. . . **Each Net** 3/6
Drapery and Drapers' Accounts. By RICHARD BEYNON.
Grocery and Grocers' Accounts. By W. F. TUPMAN.
Ironmongery and Ironmongers' Accounts. By S. W. FRANCIS.

COMMON COMMODITIES OF COMMERCE AND INDUSTRIES

In each of the handbooks in this series a particular product or industry is treated by an expert writer and practical man of business. Beginning with the life history of the plant, or other natural product, he follows its development until it becomes a commercial commodity, and so on through the various phases of its sale in the market and its purchase by the consumer.

Each book in crown 8vo, cloth, with many illustrations, 2s. 6d. net.

TEA
COFFEE
SUGAR
OILS
WHEAT AND ITS PRODUCTS
RUBBER
IRON AND STEEL
COPPER
COAL
TIMBER
LEATHER
COTTON
SILK
WOOL
LINEN
TOBACCO
CLAYS AND CLAY PRODUCTS
PAPER
SOAP
GLASS AND GLASS MAKING
GUMS AND RESINS
THE MOTOR INDUSTRY
THE BOOT AND SHOE INDUSTRY

GAS AND GAS MAKING
FURNITURE
COAL TAR AND SOME OF ITS PRODUCTS
PETROLEUM
SALT AND THE SALT INDUSTRY
KNITTED FABRICS
ZINC
CORDAGE AND CORDAGE HEMP AND FIBRES
CARPETS
ASBESTOS
PHOTOGRAPHY
ACIDS AND ALKALIS
SILVER
GOLD
PAINTS AND VARNISHES
ELECTRICITY
ALUMINIUM
BUTTER AND CHEESE
BRITISH CORN TRADE
ENGRAVING
LEAD
STONES AND QUARRIES

LAW

THE ELEMENTS OF COMMERCIAL LAW. By A. H. DOUGLAS, LL.B. (Lond.). In crown 8vo, cloth, 128 pp. . . . . **Net** 2/-
THE COMMERCIAL LAW OF ENGLAND. By J. A. SLATER, B.A., LL.B. (Lond.). In crown 8vo, cloth, 252 pp. Seventh Edition . . . **Net** 3/6
THE LAW OF CONTRACT. By R. W. HOLLAND, M.A., M.Sc., LL.D. *Of the Middle Temple, Barrister-at-Law.* In foolscap 8vo, cloth, 120 pp. . . **Net** 1/6
QUESTIONS AND ANSWERS IN COMMERCIAL LAW. By J. WELLS THATCHER, *Barrister-at-Law.* In crown 8vo, cloth gilt, 172 pp. . . **Net** 2/6
EXAMINATION NOTES ON COMMERCIAL LAW. By R. W. HOLLAND, O.B.E., M.A., M.Sc., LL.D. Cloth, 6½ in. by 3½ in., 56 pp. . . . **Net** 1/-
NOTES ON BANKING AND COMMERCIAL LAW. By T. LLOYD DAVIES. In foolscap 8vo, 100 pp. . . . . **Net** 3/-
ELEMENTARY LAW. By E. A. COPE. In crown 8vo, cloth, 228 pp. . **Net** 2/6
LEGAL TERMS, PHRASES, AND ABBREVIATIONS. By E. A. COPE. Third Edition. In crown 8vo, cloth, 216 pp. . . . . . **Net** 3/-
SOLICITOR'S CLERK'S GUIDE. By the same Author. In crown 8vo, cloth gilt, 216 pp. . . . . **Net** 3/6
CONVEYANCING. By E. A. COPE. In crown 8vo, cloth, 206 pp. . **Net** 3/6
WILLS, EXECUTORS, AND TRUSTEES. With a Chapter on Intestacy. By J. A. SLATER, B.A., LL.B. (Lond.). In foolscap 8vo, cloth, 122 pp. . **Net** 2/6
THE LAW RELATING TO TRADE CUSTOMS, MARKS, SECRETS, RESTRAINTS, AGENCIES, etc., etc. By LAWRENCE DUCKWORTH, *Barrister-at-Law.* In foolscap 8vo, cloth, 116 pp. . . . . . . **Net** 1/3
MERCANTILE LAW. By J. A. SLATER, B.A., LL.B. (Lond.). In demy 8vo, cloth gilt, 464 pp. Fourth Edition . . . . . **Net** 7/6
BILLS, CHEQUES, AND NOTES. By J. A. SLATER, B.A., LL.B. Third Edition, In demy 8vo, cloth gilt, 214 pp. . . . . . **Net** 6/-
PRINCIPLES OF MARINE LAW. By LAWRENCE DUCKWORTH. Third Edition, Revised and Enlarged. In demy 8vo, cloth gilt, 400 pp. . . **Net** 7/6
OUTLINES OF COMPANY LAW. By F. D. HEAD, B.A. (Oxon.). In demy 8vo, cloth, 100 pp. . . . . **Net** 2/-

GUIDE TO COMPANY LAW. By R. W. HOLLAND, O.B.E., M.A., M.Sc., LL.D. In crown 8vo, cloth gilt, 203 pp. . . . . . . Net 3/6
EXAMINATION NOTES ON COMPANY LAW. By R. W. HOLLAND, O.B.E., M.A., M.Sc., LL.D. Cloth, 6½ in. by 3½ in., 56 pp. . . . Net 1/-
COMPANIES AND COMPANY LAW. Together with the Companies (Consolidation) Act, 1908, and the Act of 1913. By A. C. CONNELL, LL.B. (Lond.). Second Edition, Revised. In demy 8vo, cloth gilt, 348 pp. . . Net 6/-
COMPANY CASE LAW. A digest of leading decisions. By F. D. HEAD, B.A. (Oxon.). In demy 8vo, cloth gilt, 314 pp. . . . . . Net 7/6
THE STUDENT'S GUIDE TO RAILWAY LAW. By ARTHUR E. CHAPMAN, M.A., LL.D. (Camb.). In crown 8vo, cloth gilt, 200 pp. . . Net 2/6
RAILWAY (REBATES) CASE LAW. By GEO. B. LISSENDEN. In demy 8vo, cloth gilt, 450 pp. . . . . Net 10/6
THE LAW RELATING TO SECRET COMMISSIONS AND BRIBES (CHRISTMAS BOXES, GRATUITIES, TIPS, etc.). The Prevention of Corruption Act, 1906. By ALBERT CREW, *Barrister-at-Law*. In demy 8vo, cloth gilt, 198 pp. / . Net 5/-
INHABITED HOUSE DUTY. By W. E. SNELLING. In demy 8vo, cloth gilt, 357 pp. Net 12/6
THE LAW OF CARRIAGE. By J. E. R. STEPHENS, B.A., *of the Middle Temple, Barrister-at-Law*. In demy 8vo, cloth gilt, 340 pp. . . Net 5/-
THE LAW RELATING TO THE CARRIAGE BY LAND OF PASSENGERS, ANIMALS, AND GOODS. By S. W. CLARKE, *of the Middle Temple, Barrister-at-Law*. In demy 8vo, cloth gilt, 350 pp. . . . Net 7/6
THE STUDENT'S GUIDE TO BANKRUPTCY LAW AND WINDING UP OF COMPANIES. By F. PORTER FAUSSET, B.A., LL.B., *Barrister-at-Law*. In crown 8vo, cloth gilt, 196 pp. . . . . . Net 2/6
BANKRUPTCY, DEEDS OF ARRANGEMENT AND BILLS OF SALE. By W. VALENTINE BALL, M.A., and G. MILLS, B.A., *Barristers-at-Law*. Third Edition, Revised and Enlarged. In demy 8vo, cloth gilt, 364 pp. . Net 5/-
GUIDE TO THE LAW OF LICENSING. The Handbook for all Licence Holders. By J. WELLS THATCHER. In demy 8vo, cloth gilt, 196 pp. . Net 5/-
LAW OF REPAIRS AND DILAPIDATIONS. A Handbook for Students and Practitioners. By T. CATO WORSFOLD, M.A., LL.D. In crown 8vo, cloth gilt, 104 pp. . . . . Net 3/6
THE LAW OF PROCEDURE. A Handbook for Students and Practitioners. By W. NEMBHARD HIBBERT, LL.D. In demy 8vo, cloth gilt, 122 pp. . Net 5/-
HANDBOOK OF LOCAL GOVERNMENT LAW. By J. WELLS THATCHER. In large crown 8vo, cloth gilt, 250 pp. . . . . Net 3/6
THE LAW RELATING TO THE CHILD: ITS PROTECTION, EDUCATION, AND EMPLOYMENT. By R. W. HOLLAND, O.B.E., M.A., M.Sc., LL.D. In demy 8vo, cloth gilt, 166 pp. . . . . . . Net 5/-
INCOME TAX AND SUPER-TAX LAW AND CASES. (*See* p. 8.)

FOREIGN LANGUAGES

FRENCH

A CHILD'S FIRST STEPS IN FRENCH. By A. VIZETELLY. An elementary French reader with vocabulary. Illustrated. In crown 8vo, limp cloth, 64 pp. . Net 1/3
FRENCH COURSE. Part I. In crown 8vo, 120 pp., limp cloth . Net 1/3
FRENCH COURSE. Part II. (*In preparation*)
PROGRESSIVE FRENCH GRAMMAR. By Dr. F. A. HEDGCOCK, M.A. . Net 5/6
(Also in 2 vols.: Part I, 3/6 net; Part II, 2/6 net)
Key . . . . Net 3/6
EASY FRENCH CONVERSATIONAL SENTENCES. In crown 8vo, 32 pp. . Net 6d.
ADVANCED FRENCH CONVERSATIONAL EXERCISES. In crown 8vo, 32 pp. Net 6d.
TOURISTS' VADE MECUM OF FRENCH COLLOQUIAL CONVERSATION. Handy size for the pocket, cloth . . . . . . Net 1/3
FRENCH VOCABULARIES AND IDIOMATIC PHRASES. By E. J. KEALEY, B.A. In crown 8vo, 151 pp. . . . . Net 2/-
GRADUATED LESSONS IN COMMERCIAL FRENCH. By F. MARSDEN. In crown 8vo, cloth, 150 pp. . . . . Net 2/-
FRENCH-ENGLISH AND ENGLISH-FRENCH COMMERCIAL DICTIONARY. By F. W. SMITH. In crown 8vo, cloth, 576 pp. . . . Net 7/6
COMMERCIAL FRENCH GRAMMAR. By F. W. DRAPER, M.A., B. ès L. In crown 8vo, cloth gilt, 166 pp. . . . . . Net 2/6
RAPID METHOD OF SIMPLIFIED FRENCH CONVERSATION. By V. F. HIBBERD. In crown 8vo, cloth, 192 pp. . . . . Net 2/6

GRADUATED FRENCH-ENGLISH COMMERCIAL CORRESPONDENCE. By MAURICE DENEVE. In crown 8vo, 160 pp. . . . . Net	2/-
FRENCH BUSINESS LETTERS. First Series. In crown 4to, 32 pp. . Net	8d.
FRENCH BUSINESS LETTERS. By A. H. BERNAARDT. Second Series. In crown 8vo, 48 pp. . . . . Net	8d.
COMMERCIAL CORRESPONDENCE IN FRENCH. In crown 8vo, cloth, 240 pp. Net	3/6
MERCANTILE CORRESPONDENCE. English-French. In crown 8vo, cloth 250 pp. Net	3/6
MODELS AND EXERCISES IN COMMERCIAL FRENCH. By E. T. GRIFFITHS, M.A. In crown 8vo, cloth, 180 pp. . . . . Net	2/6
FRENCH COMMERCIAL PHRASES AND ABBREVIATIONS WITH TRANSLATION. In crown 8vo, 32 pp.	6d.
FRENCH BUSINESS CONVERSATIONS AND INTERVIEWS. In crown 8vo, 80 pp., limp cloth . . . . Net	2/-
READINGS IN COMMERCIAL FRENCH. With Notes and Translations in English. In crown 8vo, cloth, 90 pp. . . . . Net	1/-
FRENCH COMMERCIAL READER. In crown 8vo, cloth, 208 pp. . . Net	2/6
ENGLISH-FRENCH AND FRENCH-ENGLISH DICTIONARY OF BUSINESS WORDS AND TERMS. Size 2 in. by 6 in., cloth, rounded corners, 540 pp. . .	4/6
FRENCH FOUNDATION BOOK OF VERBS, ACCIDENCE, AND SYNTAX. By F. A. HEDGCOCK, M.A. In crown 8vo, 90 pp. . . . . . Net	1/-
VEST POCKET LIST OF ENDINGS OF FRENCH REGULAR AND AUXILIARY VERBS. With Notes on the Participles and the Infinitive. Size 2¾ in. by 1¾ in. 48 pp. . . . . Net	2d.

GERMAN

GERMAN COURSE. Part I. 9d. net. Cloth . . . . Net	1/-
KEY TO GERMAN COURSE. In crown 8vo . . . . Net	1/6
PRACTICAL GERMAN GRAMMAR. In crown 8vo, 102 pp. . . . cloth	2/6
EASY LESSONS IN GERMAN. By J. BITHELL, M.A. In crown 8vo, cloth, 116 pp. . . . . Net	1/3
EASY GERMAN CONVERSATIONAL SENTENCES. In crown 8vo, 32 pp. . Net	8d.
ADVANCED GERMAN CONVERSATIONAL EXERCISES. In crown 8vo, 32 pp. Net	6d.
TOURISTS' VADE MECUM OF GERMAN COLLOQUIAL CONVERSATION. In crown 8vo, cloth . . . . Net	1/3
EXAMINATION NOTES ON GERMAN. By A. HARGREAVES, M.A., Ph.D. Cloth, 6¼ in by 3½ in., 56 pp. . . . . Net	1/-
GERMAN EXAMINATION PAPERS WITH MODEL ANSWERS. In crown 8vo, 48 pp. . . . . Net	6d.
COMMERCIAL GERMAN GRAMMAR. By J. BITHELL, M.A. In crown 8vo, cloth gilt, 182 pp. . . . . Net	2/6
GERMAN BUSINESS INTERVIEWS, Nos. 1 and 2. Each in crown 8vo, limp cloth. No. 1, 100 pp. ; No. 2, 74 pp. . . . . Net	1/6
ELEMENTARY GERMAN CORRESPONDENCE. By LEWIS MARSH, M.A. In crown 8vo, cloth, 143 pp. . . . . Net	2/-
COMMERCIAL CORRESPONDENCE IN GERMAN. In crown 8vo, cloth, 240 pp. Net	3/6
MERCANTILE CORRESPONDENCE. English-German. In crown 8vo, cloth, 250 pp. . . . . Net	3/6
GERMAN BUSINESS LETTERS. First Series. In crown 8vo, 48 pp. . Net	6d.
GERMAN BUSINESS LETTERS. By G. ALBERS. Second Series. In crown 8vo, 48 pp. . . . . Net	6d.
GRADUATED GERMAN-ENGLISH COMMERCIAL CORRESPONDENCE. In crown 8vo, cloth . . . . Net	3/6
GERMAN COMMERCIAL PHRASES. In crown 8vo, 32 pp. . . Net	6d.
GERMAN COMMERCIAL READER. In crown 8vo, cloth, 208 pp. . Net	3/6
READINGS IN COMMERCIAL GERMAN. With Notes and Translations in English. In crown 8vo, cloth, 90 pp. . . . . Net	1/-
ENGLISH-GERMAN AND GERMAN-ENGLISH DICTIONARY OF BUSINESS WORDS AND TERMS. Size 2 in. by 6 in., rounded corners, cloth, 440 pp. . . Net	4/6

SPANISH

EASY SPANISH CONVERSATIONAL SENTENCES. In crown 8vo, 32 pp. . Net	6d.
ADVANCED SPANISH CONVERSATIONAL EXERCISES. In crown 8vo, 32 pp. Net	6d.
TOURISTS' VADE MECUM OF SPANISH COLLOQUIAL CONVERSATION. Cloth . . . . Net	1/3
EXAMINATION NOTES ON SPANISH. By ALFRED CALVERT. Cloth, 6¼ in. by 3½ in., 56 pp. . . . . Net	1/-
COMMERCIAL SPANISH GRAMMAR. By C. A. TOLEDANO. In crown 8vo, cloth gilt, 250 pp. . . . . Net	4/6
Key . Net	2/-

SPANISH VERBS, Regular and Irregular. By G. R. MACDONALD. In crown 8vo, cloth, 180 pp. . . . . **Net**
COMMERCIAL CORRESPONDENCE IN SPANISH. In crown 8vo, cloth, 240 pp. **Net**
MANUAL OF SPANISH COMMERCIAL CORRESPONDENCE. By G. R. MACDONALD. In crown 8vo, cloth gilt, 328 pp. . . . **Net**
LESSONS IN SPANISH COMMERCIAL CORRESPONDENCE. By the same Author. In crown 8vo, cloth, 107 pp. . . . . . **Net**
SPANISH COMMERCIAL READER. By G. R. MACDONALD. In crown 8vo, cloth, 178 pp. . . . . **Net**
READINGS IN COMMERCIAL SPANISH. With Notes and Translations in English. In crown 8vo, cloth, 90 pp. . . . . . **Net**
SPANISH BUSINESS LETTERS. First Series. In crown 8vo, 32 pp. . **Net**
SPANISH BUSINESS LETTERS. By E. McCONNELL. Second Series. In crown 8vo, 48 pp. . . . . **Net**
SPANISH COMMERCIAL PHRASES. With Abbreviations and Translation. In crown 8vo, 32 pp. . . . . **Net**
SPANISH BUSINESS CONVERSATIONS AND INTERVIEWS. With Correspondence Invoices, etc. In crown 8vo, 90 pp, limp cloth . . . **Net**
SPANISH-ENGLISH AND ENGLISH-SPANISH COMMERCIAL DICTIONARY. By G. R. MACDONALD. In crown 8vo, cloth gilt, 652 pp. . . **Net**
COMMERCIAL AND TECHNICAL TERMS IN ENGLISH AND SPANISH. By R. D. MONTEVERDE, B.A. In crown 8vo, ... **Net**
SPANISH IDIOMS, with their English Equivalents. By the same Author. In crown 8vo . . . . **Net**

ITALIAN

TOURISTS' VADE MECUM OF ITALIAN COLLOQUIAL CONVERSATION. Cloth . . . . **Net**
COMMERCIAL ITALIAN GRAMMAR. By LUIGI RICCI. In crown 8vo, cloth gilt, 154 pp. . . . . **Net**
MERCANTILE CORRESPONDENCE. English-Italian. In crown 8vo, cloth, 250 pp. . . . . **Net**
ITALIAN BUSINESS LETTERS. By A. VALGIMIGLI. In crown 8vo, 48 pp. . **Net**
BARETTI'S DICTIONARY OF THE ITALIAN AND ENGLISH LANGUAGES. By J. DAVENPORT and G. COMELATI. Two volumes. In demy 8vo, cloth gilt, about 1,500 pp. . . . . **Net**

MISCELLANEOUS

PRACTICAL PORTUGUESE GRAMMAR, By C. A. and A. TOLEDANO. In crown 8vo, cloth, 330 pp. . . . . **Net**
MERCANTILE CORRESPONDENCE. English-Portuguese. In crown 8vo, cloth, 250 pp. . . . . **Net**
LESSONS IN PORTUGUESE COMMERCIAL CORRESPONDENCE. By G. R. MACDONALD. In crown 8vo, cloth, 108 pp. . . . **Net**
DICTIONARY OF COMMERCIAL CORRESPONDENCE IN ENGLISH, FRENCH, GERMAN, SPANISH, ITALIAN, PORTUGUESE, AND RUSSIAN. Third Revised Edition. In demy 8vo, cloth, 718 pp. . . . **Net**
THE FOREIGN CORRESPONDENT. By EMIL DAVIES. In crown 8vo, cloth, 80 pp. . . . . **Net**
COMMERCIAL TERMS IN FIVE LANGUAGES. Being about 1,900 terms and phrases used in commerce, with their equivalents in French, German, Spanish, and Italian. Size 3 in. by 4¾ in., cloth, 118 pp. . . . **Net**
A NEW DICTIONARY OF THE PORTUGUESE AND ENGLISH LANGUAGES. Based on a manuscript of Julius Cornet, by H. MICHAELIS. In two parts, demy 8vo cloth gilt, 1,478 pp. . . . . . . **Each, Net**
Abridged Edition, 783 pp. (two parts in one volume) . . **Net**
INTERNATIONAL TECHNICAL DICTIONARY IN ENGLISH, ITALIAN, FRENCH, AND GERMAN. By E. WEBBER. In foolscap 16mo., 921 pp., cloth . **Net**

PITMAN'S SHORTHAND

All books are in foolscap 8vo size unless otherwise stated.

INSTRUCTION BOOKS

Centenary Editions.

PITMAN'S SHORTHAND TEACHER. An elementary work suited for self-instruction or class teaching
KEY TO "PITMAN'S SHORTHAND TEACHER"

PITMAN'S SHORTHAND PRIMERS. In three Books: Elementary, Intermediate, and Advanced . . . . Each, 8d. Keys, each	8d.
PITMAN'S SHORTHAND READING LESSONS. Nos. 1, 2 and 3 . . Each	8d.
KEYS TO "PITMAN'S SHORTHAND READING LESSONS," Nos, 1, 2, and 3 Each	3d.
PITMAN'S SHORTHAND COPY BOOKS. Nos. 1, 2, 3, and 4. An entirely new series covering the theory of the system. Foolscap 4to (8¾ in. by 6½ in.) . Each	6d.
PITMAN'S SHORTHAND DRILL EXERCISES. Oblong	8d.
COMPEND OF PITMAN'S SHORTHAND.	2d.
PITMAN'S SHORTHAND INSTRUCTOR. Complete Instruction in the System. Cloth	4/-
Key, 2/-; cloth	2/6
THE CENTENARY CHANGES IN PITMAN'S SHORTHAND. In crown 8vo .	1d.
SUMMARIES FROM "PITMAN'S SHORTHAND INSTRUCTOR." Size, 2⅞ in. by 4 in.	3d.
PITMAN'S SHORTHAND MANUAL. Contains instruction in the Intermediate Style, with 100 Exercises. 2/-. Cloth 2/6 . . . . Key	8d.
PITMAN'S SHORTHAND GRADUS. Writing Exercises in ordinary print for *Manual*	3d.
PITMAN'S SHORTHAND REPORTER. Containing instruction in the Advanced Style: with 52 Exercises. 2/6. Cloth 3/- . . . . . . Key	8d.
REPORTING EXERCISES. Exercises on all the rules and contracted words. In ordinary print, counted for dictation . . . . . . 6d.; Key	1/-
PITMAN'S SHORTHAND CATECHISM. In crown 8vo	1/6
PITMAN'S SHORTHAND WRITING EXERCISES AND EXAMINATION TESTS. In crown 8vo, paper boards. 2/- . . . . Key	3/6
EXAMINATION NOTES ON PITMAN'S SHORTHAND. By H. W. B. WILSON. 8 in. by 3½ in., cloth	1/6
GRADED SHORTHAND READINGS.	
Elementary, with Key. In crown 8vo, oblong	6d.
Intermediate, with Key. In crown 8vo, oblong	8d.
Second Series	8d.
Advanced, with Key. In crown 8vo, oblong	8d.
GRADUATED TESTS IN PITMAN'S SHORTHAND. Illustrating all the rules in the Intermediate Style. In note-book form, post 8vo (6½ in. by 4¼ in.), with ruled paper	8d.
PROGRESSIVE STUDIES IN PITMAN'S SHORTHAND.	1/-
TALKS WITH SHORTHAND STUDENTS. By JAMES HYNES . . .	1/-
CHATS ABOUT PITMAN'S SHORTHAND. By GEORGE BLETCHER . .	1/-
LECTURETTES ON PITMAN'S SHORTHAND. By J. HYNES . . .	1/-
PITMAN'S SHORTHAND RAPID COURSE. A Series of Twenty Simple Lessons covering the whole of the system and specially adapted for business purposes. In crown 8vo. Cloth 2/6 . . . Key 2/6 With Additional Exercises	4/-
PITMAN'S SHORTHAND RAPID COURSE, ADDITIONAL EXERCISES ON .	8d.
READING EXERCISES ON THE RAPID COURSE (In Shorthand), crown 8vo, 62 pp.	1/-
PITMAN'S SHORTHAND COMMERCIAL COURSE. Specially adapted for commercial students, Cloth 4/- . . . . Key, 2/6; Additional Exercises	1/-
PITMAN'S EXERCISES IN BUSINESS SHORTHAND. By A. BENJAMIN, I.P.S. (Hons.), F.C.Sp.T.	1/9

GRAMMALOGUES AND CONTRACTIONS

GRAMMALOGUES AND CONTRACTIONS. For use in classes . .	2d.
VEST POCKET LIST OF GRAMMALOGUES AND CONTRACTIONS OF PITMAN'S SHORTHAND. 2¾ in. by 1⅞ in., limp cloth	2d.
EXERCISES ON THE GRAMMALOGUES AND CONTRACTIONS OF PITMAN'S SHORTHAND. By J. F. C. GROW. In Shorthand, with Key. In crown 8vo, limp cloth	6d.
HOW TO PRACTISE AND MEMORIZE THE GRAMMALOGUES OF PITMAN'S SHORTHAND. Compiled by D. J. GEORGE. Size 7¾ in. by 5 in. . .	4d.

SHORTHAND DICTIONARIES

PITMAN'S ENGLISH AND SHORTHAND DICTIONARY. In crown 8vo, cloth, 820 pp.	7/6
PITMAN'S SHORTHAND DICTIONARY. Crown 8vo (7¼ in. by 5⅛ in.), 378 pp. Cloth	6/-
PITMAN'S POCKET SHORTHAND DICTIONARY. Royal 32mo (3⅝ in. by 4¾ in.). Cloth	2/-
PITMAN'S REPORTER'S ASSISTANT. In crown 8vo, cloth . . .	3/6

SHORTHAND PHRASE BOOKS, ETC.

PHONOGRAPHIC PHRASE BOOK. . . . . . 1/6; Cloth	2/-
SHORTHAND WRITERS' PHRASE BOOKS AND GUIDES. Each in foolscap 8vo, Cloth . . . . Net	2/-

Electrical and Engineering, Railway, Estate Agents, etc., Printing and Publishing, Insurance, Banking, Stockbroking and Financial, Commercial, Legal, Municipal, Builders and Contractors, Shipping, Iron and Steel Trades, Civil Engineering, Naval and Military, Chemical and Drug, Provision Trades.
MEDICAL REPORTING IN PITMAN'S SHORTHAND. By H. DICKINSON. With an Introduction and Lists of Phraseograms, Outlines, and Abbreviations. In crown 8vo, cloth . . . . **Net**
SHORTHAND CLERK'S GUIDE. By VINCENT E. COLLINGE, A.C.I.S. In crown 8vo, cloth . . . .

DICTATION AND SPEED PRACTICE BOOKS

SPECIALISED CORRESPONDENCE BOOKS. (1) The Chemical Trade. (2) The Paper Trade. (3) The Building Trade. In ordinary print . . . **Each**
GRADUATED DICTATION BOOKS. (1) Political Speeches. (2) Sermons. In ordinary print. In crown 8vo. . . . . **Each**
STUDENT'S PRACTICE BOOK. In cr. 8vo, 241 pp. . . . .
GRADUATED DICTATION BOOKS. (New Series) I and II. . **Each**
GRADUATED COMMERCIAL LETTERS FOR DICTATION. 8¼ in. by 6 in. .
REPORTING PRACTICE. In crown 8vo, cloth . . . .
PROGRESSIVE DICTATOR. Third Edition. In crown 8vo, cloth
SHORTHAND CANDIDATE'S DICTATION EXERCISES. In crown 8vo, cloth
COMMERCIAL DICTATION AND TYPEWRITING
SPEED TESTS AND GUIDE TO RAPID WRITING IN SHORTHAND. In crown 8vo
FIVE MINUTE SPEED TESTS. With Introduction on Acquisition of Speed by P. P. JACKSON. In crown 8vo,
CUMULATIVE SPELLER AND SHORTHAND VOCABULARY. By CHARLES E. SMITH. In crown 8vo, paper boards
POCKET DICTATION BOOKS, Nos. 1, 2, 3, and 4. 2⅝ in. by 3¾ in. . **Each**
SPEED TRAINING IN PITMAN'S SHORTHAND. By T. F. MARRINER
ACQUISITION OF SPEED IN SHORTHAND. By E. A. COPE. In ordinary print. In crown 8vo
BROWN'S SHORT-CUTS IN SHORTHAND. By GEORGE BROWN, F.I.P.S. In crown 8vo. . . . .
THE STENOGRAPHIC EXPERT. By W. B. BOTTOME and W. F. SMART. In demy 8vo, cloth . . . . **Net**
SHORTHAND COMMERCIAL LETTER-WRITER. Advanced Style . **1/-; Key**
OFFICE WORK IN SHORTHAND. Specimens of Legal and other Professional Work commonly dictated to Shorthand clerks, in the Advanced Style **1/3; Key**
COMMERCIAL CORRESPONDENCE IN SHORTHAND. In crown 8vo, cloth
BUSINESS CORRESPONDENCE IN SHORTHAND. In the Advanced Style. **1/3; Key**
TRADE CORRESPONDENCE IN SHORTHAND. In the Advanced Style. **1/3; Key**
MISCELLANEOUS CORRESPONDENCE IN PITMAN'S SHORTHAND. First, Second, Third, and Fourth Series. Advanced Style, with Keys in ordinary print. Each in crown 8vo, oblong . . . . **Net**

SHORTHAND READING BOOKS

In the Elementary Style.
AESOP'S FABLES . . . .
EASY READINGS. With Key . . . .
LEARNER'S SHORTHAND READER. Illustrated. . . . .
STIRRING TALES . . . .
PERILS OF THE BUSH AND OTHER STORIES . . . .
In the Intermediate Style.
PITMAN'S PHONOGRAPHIC READER, No. 1. With Key . . .
GULLIVER'S VOYAGE TO LILLIPUT. By JONATHAN SWIFT. With Key. Cloth
SUBMARINE X7 AND OTHER STORIES. Illustrated . . .
THE VICAR OF WAKEFIELD. By OLIVER GOLDSMITH. Illustrated. **2/-;** Cloth
TALES AND SKETCHES. By WASHINGTON IRVING. With Key. **1/6;** Cloth
TALES OF ADVENTURE. By various Authors . . . .
THE RUNAWAY AIRSHIP AND OTHER STORIES. . . .
THE SILVER SHIP OF MEXICO. An abridgment of J. H. INGRAHAM's Story Cloth
SELECT READINGS . . . . . . **No. 1, 6d. No. II**
THE BOOK OF PSALMS. Bible Authorised Version. Cloth gilt, red edges .
COMMERCIAL READERS IN SHORTHAND. (1) Commercial Institutions, 8d. (2) Commodities. (3) Leaders of Commerce. (4) Gateways of British Commerce. **Each**

In the Advanced Style.

PHONOGRAPHIC READER II. With Key	6d.
A CHRISTMAS CAROL. By CHARLES DICKENS. .1/3; Cloth	1/9
TALES FROM DICKENS . . . . Cloth	2/6
THE SIGN OF FOUR. By SIR A. CONAN DOYLE . Cloth	2/-
THE RETURN OF SHERLOCK HOLMES. Vols. I, II and III Each, cloth	2/-
AROUND THE WORLD IN EIGHTY DAYS. By JULES VERNE	2/-
SELF-CULTURE. By J. S. BLACKIE. . 1/-; Cloth, 1/6; Key	2/6
SELECTIONS FROM AMERICAN AUTHORS. With Key	1/3
THE LEGEND OF SLEEPY HOLLOW. By WASHINGTON IRVING. With Key	8d.
RIP VAN WINKLE. By WASHINGTON IRVING. With Key	6d.
A COURSE IN BUSINESS TRAINING. By G. K. BUCKNALL, A.C.I.S. (Shorthand Edition), 288 pp.	3/-

SHORTHAND TEACHERS' BOOKS

PITMAN'S SHORTHAND TEACHER'S HANDBOOK. In crown 8vo, cloth	1/6
NOTES OF LESSONS ON PITMAN'S SHORTHAND. Size 8 in. by 3¾ in., cloth	2/6
PREPARATION FOR A SHORTHAND TEACHER'S EXAMINATION. Size 8 in. by 3¾ in., cloth	1/6
A COMMENTARY ON PITMAN'S SHORTHAND. By J. W. TAYLOR. In foolscap 8vo, cloth gilt, 448 pp.	5/-
THE METHODS OF TEACHING SHORTHAND. By E. J. MCNAMARA, M.A. In crown 8vo, cloth	2/6
CHART OF THE PHONOGRAPHIC ALPHABET. 22 in. by 35 in.	2d.
CHARTS ON PITMAN'S SHORTHAND. Twenty large Charts (22 in. by 35 in.) The Set	7/6
DERIVATIVE AND COMPOUND WORDS IN PITMAN'S SHORTHAND By H. W. B. WILSON. In foolscap 8vo	2/-
HISTORY OF SHORTHAND. By SIR ISAAC PITMAN. Fourth Edition, Revised. In crown 8vo, cloth . . . . Net	5/-

TYPEWRITING

THE JUNIOR TYPIST. By ANNIE E. DAVIS. Demy 8vo, cloth . Net	1/6
NEW COURSE IN TYPEWRITING. By MRS. SMITH CLOUGH. Large post 4to	2/-
PITMAN'S TYPEWRITER MANUAL. Can be used with any machine. Sixth Edition. Large post 4to, cloth	5/6
PITMAN'S TYPEWRITING EXAMPLES for any machine—	
On cards, 48 examples, foolscap folio	3/-
In oblong note-book, for standing by the side of the machine	2/6
In note-book form, in covers	2/-
PITMAN'S EXERCISES AND TESTS IN TYPEWRITING. Foolscap folio. Quarter cloth. Third Edition, revised	4/-
HOW TO TEACH TYPEWRITING. By KATE PICKARD, B.A. (Lond.). Crown 4to. cloth . . . . Net	5/-
PRACTICAL COURSE IN TOUCH TYPEWRITING. By C. E. SMITH. English Edition, revised and enlarged. Size, 8½ in. by 11 in.	2/6
PRACTICAL TOUCH TYPEWRITING CHART. Size, 30 in. by 40 in. . Net	2/6
REMINGTON TYPEWRITER MANUAL. For Nos. 5 and 7, 10 and 11. With Exercises and illustrations. Ninth Edition. Large post 4to . . Net	2/-
THE UNDERWOOD TYPEWRITER MANUAL. By A. J. SYLVESTER. Large post 4to	2/6
BAR-LOCK TYPEWRITER MANUAL (Group System of Touch Typewriting). By H. ETHERIDGE. Large post 4to . . . . . Net	3/-
ROYAL SOCIETY OF ARTS TYPEWRITING TESTS. By A. E. MORTON. Elem., Inter., and Advanced. Each in foolscap folio . . . Net	4/-
MODERN TYPEWRITING AND MANUAL OF OFFICE PROCEDURE. By A. E. MORTON. 6½ in. by 9½ in., cloth	5/6
A TYPEWRITING CATECHISM. By MRS. SMITH CLOUGH. In large post 4to Net	4/-
DICTIONARY OF TYPEWRITING. By H. ETHERIDGE. In demy 8vo, cloth, fully illustrated . . . . Net	6/-
HIGH SPEED IN TYPEWRITING. By A. M. KENNEDY and F. JARRETT. In demy 4to, 72 pp.	2/6

PERIODICALS

PITMAN'S JOURNAL. Subscription, which may begin at any time, **17/4** per annum, post free. (Estab. 1842.) 24 pp. . . Weekly **3d.**, by post	4d.
PITMAN'S SHORTHAND WEEKLY. (Estab. 1892.) . Weekly **2d.**, by post	2½d.
BUSINESS ORGANISATION AND MANAGEMENT. Monthly. **1/6** net, by post **1/9**. Annual Subscription . . . . . . Net	18/-

---

Pitman's Complete Commercial and Shorthand Catalogues containing FULL *particulars of these and other important works will be sent post free on application.*

This volume from the
Cornell University Library's
print collections was scanned on an
APT BookScan and converted
to JPEG 2000 format
by Kirtas Technologies, Inc.,
Victor, New York.
Color images scanned as 300 dpi
(uninterpolated), 24 bit image capture
and grayscale/bitonal scanned
at 300 dpi 24 bit color images
and converted to 300 dpi
(uninterpolated), 8 bit image capture.
All titles scanned cover to
cover and pages may include
marks, notations and other
marginalia present in the
original volume.

The original volume was digitized
with the generous support of the
Microsoft Corporation
in cooperation with the
Cornell University Library.

Cover design by Lou Robinson,
Nightwood Design.

Printed in Great Britain
by Amazon